e

The Story of a Number

e

The Story of a Number

Eli Maor

PRINCETON UNIVERSITY PRESS

PRINCETON, NEW JERSEY

2
2.7
2.71
2.718
2.71 82
2.71 828
2.71 8281
2.71 82818
2.71 828182
2.71 8281828
2.71 82818284

Copyright © 1994 by Princeton University Press
Published by Princeton University Press, 41 William Street,
Princeton, New Jersey 08540
In the United Kingdom: Princeton University Press,
Chichester, West Sussex

Library of Congress Cataloging-in-Publication Data

Maor, Eli.
e: the story of a number / Eli Maor.
p.   cm.
Includes bibliographical references and index.
ISBN 0-691-03390-0
1. e (The number)   I. Title.
QA247.5.M33   1994
512'.73—dc20   93-39003

This book has been composed in Adobe Times Roman

Princeton University Press books are printed
on acid-free paper and meet the guidelines
for permanence and durability of the Committee
on Production Guidelines for Book Longevity
of the Council on Library Resources

Printed in the United States of America

10   9   8   7   6   5   4   3

In memory of my parents, Richard and Luise Metzger

Philosophy is written in this grand book—I mean the universe—which stands continually open to our gaze, but it cannot be understood unless one first learns to comprehend the language and interpret the characters in which it is written. It is written in the language of mathematics, and its characters are triangles, circles, and other geometric figures, without which it is humanly impossible to understand a single word of it.

—GALILEO GALILEI, *Il Saggiatore* (1623)

Contents

Preface

It must have been at the age of nine or ten when I first encountered the number π. My father had a friend who owned a workshop, and one day I was invited to visit the place. The room was filled with tools and machines, and a heavy oily smell hung over the place. Hardware had never particularly interested me, and the owner must have sensed my boredom when he took me aside to one of the bigger machines that had several flywheels attached to it. He explained that no matter how large or small a wheel is, there is always a fixed ratio between its circumference and its diameter, and this ratio is about $3^1/_7$. I was intrigued by this strange number, and my amazement was heightened when my host added that no one had yet written this number exactly—one could only approximate it. Yet so important is this number that a special symbol has been given to it, the Greek letter π. Why, I asked myself, would a shape as simple as a circle have such a strange number associated with it? Little did I know that the very same number had intrigued scientists for nearly four thousand years, and that some questions about it have not been answered even today.

Several years later, as a high school junior studying algebra, I became intrigued by a second strange number. The study of logarithms was an important part of the curriculum, and in those days—well before the appearance of hand-held calculators—the use of logarithmic tables was a must for anyone wishing to study higher mathematics. How dreaded were these tables, with their green cover, issued by the Israeli Ministry of Education! You got bored to death doing hundreds of drill exercises and hoping that you didn't skip a row or look up the wrong column. The logarithms we used were called "common"—they used the base 10, quite naturally. But the tables also had a page called "natural logarithms." When I inquired how anything can be more "natural" than logarithms to the base 10, my teacher answered that there is a special number, denoted by the letter e and approximately equal to 2.71828, that is used as a base in "higher" mathematics. Why this strange number? I had to wait until my senior year, when we took up the calculus, to find out.

In the meantime π had a cousin of sorts, and a comparison between the two was inevitable—all the more so since their values are so close. It took me a few more years of university studies to learn that the two cousins are indeed closely related and that their relationship is all the more mysterious by the presence of a third symbol, i, the celebrated "imaginary unit," the square root of -1. So here were all the elements of a mathematical drama waiting to be told.

The story of π has been extensively told, no doubt because its history goes back to ancient times, but also because much of it can be grasped without a knowledge of advanced mathematics. Perhaps no book did better than Petr Beckmann's *A History of π*, a model of popular yet clear and precise exposition. The number e fared less well. Not only is it of more modern vintage, but its history is closely associated with the calculus, the subject that is traditionally regarded as the gate to "higher" mathematics. To the best of my knowledge, a book on the history of e comparable to Beckmann's has not yet appeared. I hope that the present book will fill this gap.

My goal is to tell the story of e on a level accessible to readers with only a modest background in mathematics. I have minimized the use of mathematics in the text itself, delegating several proofs and derivations to the appendixes. Also, I have allowed myself to digress from the main subject on occasion to explore some side issues of historical interest. These include biographical sketches of the many figures who played a role in the history of e, some of whom are rarely mentioned in textbooks. Above all, I want to show the great variety of phenomena—from physics and biology to art and music—that are related to the exponential function e^x, making it a subject of interest in fields well beyond mathematics.

On several occasions I have departed from the traditional way that certain topics are presented in calculus textbooks. For example, in showing that the function $y = e^x$ is equal to its own derivative, most textbooks first derive the formula $d(\ln x)/dx = 1/x$, a long process in itself. Only then, after invoking the rule for the derivative of the inverse function, is the desired result obtained. I have always felt that this is an unnecessarily long process: one can derive the formula $d(e^x)/dx = e^x$ directly—and much faster—by showing that the derivative of the general exponential function $y = b^x$ is proportional to b^x and then finding the value of b for which the proportionality constant is equal to 1 (this derivation is given in Appendix 4). For the expression $\cos x + i \sin x$, which appears so frequently in higher mathematics, I have used the concise notation cis x (pronounced "ciss x"), with the hope that this much shorter notation will be used more often. When considering the analogies between the circular and the hyperbolic functions, one of the most beautiful results, discovered around 1750 by Vincenzo Riccati, is that for both types of functions the independent variable can be interpreted geometrically as an area, making the formal similarities between the two types of functions even more striking. This fact—seldom mentioned in the textbooks—is discussed in Chapter 12 and again in Appendix 7.

In the course of my research, one fact became immediately clear: the number e was known to mathematicians at least half a century before the invention of the calculus (it is already referred to in Edward Wright's English translation of John Napier's work on logarithms, published in 1618). How could this be? One possible explanation is that the number e first appeared in connection with the formula for compound interest. Someone—we don't know who or when—must have noticed the curious fact that if a principal P is compounded n times a year for t years at an annual interest rate r, and if n is allowed to increase without bound, the amount of money S, as found from the formula $S = P(1 + r/n)^{nt}$, seems to approach a certain limit. This limit, for $P = 1$, $r = 1$, and $t = 1$, is about 2.718. This discovery—most likely an experimental observation rather than the result of rigorous mathematical deduction—must have startled mathematicians of the early seventeenth century, to whom the limit concept was not yet known. Thus, the very origins of the the number e and the exponential function e^x may well be found in a mundane problem: the way money grows with time. We shall see, however, that other questions—notably the area under the hyperbola $y = 1/x$—led independently to the same number, leaving the exact origin of e shrouded in mystery. The much more familiar role of e as the "natural" base of logarithms had to wait until Leonhard Euler's work in the first half of the eighteenth century gave the exponential function the central role it plays in the calculus.

I have made every attempt to provide names and dates as accurately as possible, although the sources often give conflicting information, particularly on the priority of certain discoveries. The early seventeenth century was a period of unprecedented mathematical activity, and often several scientists, unaware of each other's work, developed similar ideas and arrived at similar results around the same time. The practice of publishing one's results in a scientific journal was not yet widely known, so some of the greatest discoveries of the time were communicated to the world in the form of letters, pamphlets, or books in limited circulation, making it difficult to determine who first found this fact or that. This unfortunate state of affairs reached a climax in the bitter priority dispute over the invention of the calculus, an event that pitted some of the best minds of the time against one another and was in no small measure responsible for the slowdown of mathematics in England for nearly a century after Newton.

As one who has taught mathematics at all levels of university instruction, I am well aware of the negative attitude of so many students toward the subject. There are many reasons for this, one of them no doubt being the esoteric, dry way in which we teach the subject. We tend to overwhelm our students with formulas, definitions, theorems, and proofs, but we seldom mention the historical evolution of these facts, leaving the impression that these facts were handed to us, like the Ten Commandments, by some divine authority. The history of mathematics is a good way to correct these impressions. In my classes I always try to interject some morsels of mathematical history or vignettes of the persons whose names are associated with the formulas and theorems. The present book derives partially from this approach. I hope it will fulfill its intended goal.

Many thanks go to my wife, Dalia, for her invaluable help and support in getting this book written, and to my son Eyal for drawing the illustrations. Without them this book would never have become a reality.

Skokie, Illinois
January 7, 1993

e

The Story of a Number

1

John Napier, 1614

Seeing there is nothing that is so troublesome to
mathematical practice, nor that doth more molest and
hinder calculators, than the multiplications, divisions,
square and cubical extractions of great numbers. . . .
I began therefore to consider in my mind by what certain
and ready art I might remove those hindrances.
—JOHN NAPIER, *Mirifici logarithmorum canonis*
descriptio (1614)[1]

Rarely in the history of science has an abstract mathematical idea
been received more enthusiastically by the entire scientific community than the invention of logarithms. And one can hardly imagine a
less likely person to have made that invention. His name was John
Napier.[2]

The son of Sir Archibald Napier and his first wife, Janet Bothwell,
John was born in 1550 (the exact date is unknown) at his family's
estate, Merchiston Castle, near Edinburgh, Scotland. Details of his
early life are sketchy. At the age of thirteen he was sent to the University of St. Andrews, where he studied religion. After a sojourn abroad
he returned to his homeland in 1571 and married Elizabeth Stirling,
with whom he had two children. Following his wife's death in 1579,
he married Agnes Chisholm, and they had ten more children. The
second son from this marriage, Robert, would later be his father's
literary executor. After the death of Sir Archibald in 1608, John returned to Merchiston, where, as the eighth laird of the castle, he spent
the rest of his life.[3]

Napier's early pursuits hardly hinted at future mathematical creativity. His main interests were in religion, or rather in religious activism. A fervent Protestant and staunch opponent of the papacy, he
published his views in *A Plaine Discovery of the whole Revelation of*
Saint John (1593), a book in which he bitterly attacked the Catholic
church, claiming that the pope was the Antichrist and urging the

Scottish king James VI (later to become King James I of England) to purge his house and court of all "Papists, Atheists, and Newtrals."[4] He also predicted that the Day of Judgment would fall between 1688 and 1700. The book was translated into several languages and ran through twenty-one editions (ten of which appeared during his lifetime), making Napier confident that his name in history—or what little of it might be left—was secured.

Napier's interests, however, were not confined to religion. As a landowner concerned to improve his crops and cattle, he experimented with various manures and salts to fertilize the soil. In 1579 he invented a hydraulic screw for controlling the water level in coal pits. He also showed a keen interest in military affairs, no doubt being caught up in the general fear that King Philip II of Spain was about to invade England. He devised plans for building huge mirrors that could set enemy ships ablaze, reminiscent of Archimedes' plans for the defense of Syracuse eighteen hundred years earlier. He envisioned an artillery piece that could "clear a field of four miles circumference of all living creatures exceeding a foot of height," a chariot with "a moving mouth of mettle" that would "scatter destruction on all sides," and even a device for "sayling under water, with divers and other stratagems for harming of the enemyes"—all forerunners of modern military technology.[5] It is not known whether any of these machines was actually built.

As often happens with men of such diverse interests, Napier became the subject of many stories. He seems to have been a quarrelsome type, often becoming involved in disputes with his neighbors and tenants. According to one story, Napier became irritated by a neighbor's pigeons, which descended on his property and ate his grain. Warned by Napier that if he would not stop the pigeons they would be caught, the neighbor contemptuously ignored the advice, saying that Napier was free to catch the pigeons if he wanted. The next day the neighbor found his pigeons lying half-dead on Napier's lawn. Napier had simply soaked his grain with a strong spirit so that the birds became drunk and could barely move. According to another story, Napier believed that one of his servants was stealing some of his belongings. He announced that his black rooster would identify the transgressor. The servants were ordered into a dark room, where each was asked to pat the rooster on its back. Unknown to the servants, Napier had coated the bird with a layer of lampblack. On leaving the room, each servant was asked to show his hands; the guilty servant, fearing to touch the rooster, turned out to have clean hands, thus betraying his guilt.[6]

All these activities, including Napier's fervent religious campaigns, have long since been forgotten. If Napier's name is secure in history, it is not because of his best-selling book or his mechanical ingenuity but because of an abstract mathematical idea that took him twenty years to develop: logarithms.

The sixteenth and early seventeenth centuries saw an enormous expansion of scientific knowledge in every field. Geography, physics, and astronomy, freed at last from ancient dogmas, rapidly changed man's perception of the universe. Copernicus's heliocentric system, after struggling for nearly a century against the dictums of the Church, finally began to find acceptance. Magellan's circumnavigation of the globe in 1521 heralded a new era of marine exploration that left hardly a corner of the world unvisited. In 1569 Gerhard Mercator published his celebrated new world map, an event that had a decisive impact on the art of navigation. In Italy Galileo Galilei was laying the foundations of the science of mechanics, and in Germany Johannes Kepler formulated his three laws of planetary motion, freeing astronomy once and for all from the geocentric universe of the Greeks. These developments involved an ever increasing amount of numerical data, forcing scientists to spend much of their time doing tedious numerical computations. The times called for an invention that would free scientists once and for all from this burden. Napier took up the challenge.

We have no account of how Napier first stumbled upon the idea that would ultimatley result in his invention. He was well versed in trigonometry and no doubt was familiar with the formula

$$\sin A \cdot \sin B = 1/2[\cos(A - B) - \cos(A + B)]$$

This formula, and similar ones for $\cos A \cdot \cos B$ and $\sin A \cdot \cos B$, were known as the *prosthaphaeretic rules*, from the Greek word meaning "addition and subtraction." Their importance lay in the fact that the product of two trigonometric expressions such as $\sin A \cdot \sin B$ could be computed by finding the sum or difference of other trigonometric expressions, in this case $\cos(A - B)$ and $\cos(A + B)$. Since it is easier to add and subtract than to multiply and divide, these formulas provide a primitive system of reduction from one arithmetic operation to another, simpler one. It was probably this idea that put Napier on the right track.

A second, more straightforward idea involved the terms of a *geometric progression*, a sequence of numbers with a fixed ratio between successive terms. For example, the sequence 1, 2, 4, 8, 16, . . . is a geometric progression with the common ratio 2. If we denote the common ratio by q, then, starting with 1, the terms of the progression are 1, q, q^2, q^3, and so on (note that the nth term is q^{n-1}). Long before Napier's time, it had been noticed that there exists a simple relation between the terms of a geometric progression and the corresponding *exponents*, or indices, of the common ratio. The German mathematician Michael Stifel (1487–1567), in his book *Arithmetica integra* (1544), formulated this relation as follows: if we multiply any two terms of the progression $1, q, q^2, \ldots$, the result would be the same as if we had *added* the corresponding exponents.[7] For example, $q^2 \cdot q^3 = (q \cdot q) \cdot (q \cdot q \cdot q) = q \cdot q \cdot q \cdot q \cdot q = q^5$, a result that could have been obtained by adding the exponents 2 and 3. Similarly, dividing one term of a geometric progression by another term is equivalent to *subtracting* their exponents: $q^5/q^3 = (q \cdot q \cdot q \cdot q \cdot q)/(q \cdot q \cdot q) = q \cdot q = q^2 = q^{5-3}$. We thus have the simple rules $q^m \cdot q^n = q^{m+n}$ and $q^m/q^n = q^{m-n}$.

A problem arises, however, if the exponent of the denominator is greater than that of the numerator, as in q^3/q^5; our rule would give us $q^{3-5} = q^{-2}$, an expression that we have not defined. To get around this difficulty, we simply define q^{-n} to be $1/q^n$, so that $q^{3-5} = q^{-2} = 1/q^2$, in agreement with the result obtained by dividing q^3 by q^5 directly.[8] (Note that in order to be consistent with the rule $q^m/q^n = q^{m-n}$ when $m = n$, we must also define $q^0 = 1$.) With these definitions in mind, we can now extend a geometric progression indefinitely in both directions: $\ldots, q^{-3}, q^{-2}, q^{-1}, q^0 = 1, q, q^2, q^3, \ldots$. We see that each term is a power of the common ratio q, and that the exponents $\ldots, -3, -2, -1, 0, 1, 2, 3, \ldots$ form an *arithmetic progression* (in an arithmetic progression the *difference* between successive terms is constant, in this case 1). This relation is the key idea behind logarithms; but whereas Stifel had in mind only integral values of the exponent, Napier's idea was to extend it to a continuous range of values.

His line of thought was this: If we could write *any* positive number as a power of some given, fixed number (later to be called a base), then *multiplication and division of numbers would be equivalent to addition and subtraction of their exponents.* Furthermore, raising a number to the nth power (that is, multiplying it by itself n times) would be equivalent to *adding* the exponent n times to itself—that is, to multiplying it by n—and finding the nth root of a number would be equivalent to n repeated subtractions—that is, to division by n. In short, each arithmetic operation would be reduced to the one below it in the hierarchy of operations, thereby greatly reducing the drudgery of numerical computations.

Let us illustrate how this idea works by choosing as our base the number 2. Table 1.1 shows the successive powers of 2, beginning with $n = -3$ and ending with $n = 12$. Suppose we wish to multiply 32 by 128. We look in the table for the exponents corresponding to 32 and 128 and find them to be 5 and 7, respectively. Adding these exponents gives us 12. We now reverse the process, looking for the number whose corresponding exponent is 12; this number is 4,096, the desired answer. As a second example, suppppose we want to find 4^5. We find the exponent corresponding to 4, namely 2, and this time *multiply* it by 5 to get 10. We then look for the number whose exponent is 10 and find it to be 1,024. And, indeed, $4^5 = (2^2)^5 = 2^{10} = 1,024$.

TABLE 1.1   Powers of 2

n	-3	-2	-1	0	1	2	3	4	5	6	7	8	9	10	11	12
2^n	1/8	1/4	1/2	1	2	4	8	16	32	64	128	256	512	1,024	2,048	4,096

Of course, such an elaborate scheme is unnecessary for computing strictly with integers; the method would be of practical use only if it could be used with any numbers, integers, or fractions. But for this to happen we must first fill in the large gaps between the entries of our table. We can do this in one of two ways: by using fractional exponents, or by choosing for a base a number small enough so that its powers will grow reasonably slowly. Fractional exponents, defined by $a^{m/n} = \sqrt[n]{a^m}$ (for example, $2^{5/3} = \sqrt[3]{2^5} = \sqrt[3]{32} \approx 3.17480$), were not yet fully known in Napier's time,[9] so he had no choice but to follow the second option. But how small a base? Clearly if the base is *too* small its powers will grow too slowly, again making the system of little practical use. It seems that a number close to 1, but not too close, would be a reasonable compromise. After years of struggling with this problem, Napier decided on .9999999, or $1 - 10^{-7}$.

But why this particular choice? The answer seems to lie in Napier's concern to minimize the use of decimal fractions. Fractions in general, of course, had been used for thousands of years before Napier's time, but they were almost always written as common fractions, that is, as ratios of integers. *Decimal* fractions—the extension of our decimal numeration system to numbers less than 1—had only recently been introduced to Europe,[10] and the public still did not feel comfortable with them. To minimize their use, Napier did essentially what we do today when dividing a dollar into one hundred cents or a kilometer into one thousand meters: he divided the unit into a large number of subunits, regarding each as a new unit. Since his main goal was to reduce the enormous labor involved in trigonometric calculations, he followed the practice then used in trigonometry of dividing the radius of a unit circle into 10,000,000 or 10^7 parts. Hence, if we subtract from the full unit its 10^7th part, we get the number closest to 1 in this system, namely $1 - 10^{-7}$ or .9999999. This, then, was the common ratio ("proportion" in his words) that Napier used in constructing his table.

And now he set himself to the task of finding, by tedious repeated subtraction, the successive terms of his progression. This surely must have been one of the most uninspiring tasks to face a scientist, but Napier carried it through, spending twenty years of his life (1594–1614) to complete the job. His initial table contained just 101 entries, starting with $10^7 = 10,000,000$ and followed by $10^7(1 - 10^{-7}) = 9,999,999$, then $10^7(1 - 10^{-7})^2 = 9,999,998$, and so on up to $10^7(1 - 10^{-7})^{100} = 9,999,900$ (ignoring the fractional part .0004950), each term being obtained by subtracting from the preceding term its 10^7th part. He then repeated the process all over again, starting once more with 10^7, but this time taking as his proportion the ratio of the last number to the first in the original table, that is, 9,999,900 : 10,000,000 = .99999 or $1 - 10^{-5}$. This second table contained fifty-one entries, the last being $10^7(1 - 10^{-5})^{50}$ or very nearly 9,995,001. A third table with twenty-one entries followed, using the ratio 9,995,001 : 10,000,000; the last entry in this table was $10^7 \times .9995^{20}$, or approximately 9,900,473. Finally, from each entry in this last table Napier created sixty-eight additional entries, using the ratio 9,900,473 : 10,000,000, or very nearly .99; the last entry then turned out to be $9,900,473 \times .99^{68}$, or very nearly 4,998,609—roughly half the original number.

Today, of course, such a task would be delegated to a computer; even with a hand-held calculator the job could done in a few hours. But Napier had to do all his calculations with only paper and pen. One can therefore understand his concern to minimize the use of decimal fractions. In his own words: "In forming this progression [the entries of the second table], since the proportion between 10000000.00000, the first of the Second table, and 9995001.222927, the last of the same, is troublesome; therefore compute the twenty-one numbers in the easy proportion of 10000 to 9995, which is sufficiently near to it; the last of these, if you have not erred, will be 9900473.57808."[11]

Having completed this monumental task, it remained for Napier to christen his creation. At first he called the exponent of each power its "artificial number" but later decided on the term *logarithm*, the word meaning "ratio number." In modern notation, this amounts to saying that if (in his first table) $N = 10^7(1 - 10^{-7})^L$, then the exponent L is the (Napierian) logarithm of N. Napier's definition of logarithms differs in several respects from the modern definition (introduced in 1728 by Leonhard Euler): if $N = b^L$, where b is a fixed positive number other than 1, then L is the logarithm (to the base b) of N. Thus in Napier's system $L = 0$ corresponds to $N = 10^7$ (that is, Nap log $10^7 = 0$), whereas in the modern system $L = 0$ corresponds to $N = 1$ (that is, $\log_b 1 = 0$). Even more important, the basic rules of operation with logarithms—for example, that the logarithm of a product equals the sum of the individual logarithms—do not hold for Napier's definition. And lastly, because $1 - 10^7$ is less than 1, Napier's logarithms *decrease* with increasing numbers, whereas our common (base 10) logarithms increase. These differences are relatively minor, however, and are merely a result of Napier's insistence that the unit should be equal to 10^7 subunits. Had he not been so concerned about decimal fractions, his definition might have been simpler and closer to the modern one.[12]

In hindsight, of course, this concern was an unnecessary detour. But in making it, Napier unknowingly came within a hair's breadth of discovering a number that, a century later, would be recognized as the universal base of logarithms and that would play a role in mathematics second only to the number π. This number, e, is the limit of $(1 + 1/n)^n$ as n tends to infinity.[13]

NOTES AND SOURCES

1. As quoted in George A. Gibson, "Napier and the Invention of Logarithms," in *Handbook of the Napier Tercentenary Celebration, or Modern Instruments and Methods of Calculation*, ed. E. M. Horsburgh (1914; rpt. Los Angeles: Tomash Publishers, 1982), p. 9.

2. The name has appeared variously as Nepair, Neper, and Naipper; the correct spelling seems to be unknown. See Gibson, "Napier and the Invention of Logarithms," p. 3.

3. The family genealogy was recorded by one of John's descendants: Mark Napier, *Memoirs of John Napier of Merchiston: His Lineage, Life, and Times* (Edinburgh, 1834).

4. P. Hume Brown, "John Napier of Merchiston," in *Napier Tercentenary Memorial Volume*, ed. Cargill Gilston Knott (London: Longmans, Green and Company, 1915), p. 42.

5. Ibid., p. 47.

6. Ibid., p. 45.

7. See David Eugene Smith, "The Law of Exponents in the Works of the Sixteenth Century," in *Napier Tercentenary Memorial Volume*, p. 81.

8. Negative and fractional exponents had been suggested by some mathematicians as early as the fourteenth century, but their widespread use in mathematics is due to the English mathematician John Wallis (1616–1703) and even more so to Newton, who suggested the modern notations a^{-n} and $a^{m/n}$ in 1676. See Florian Cajori, *A History of Mathematical Notations*, vol. 1, *Elementary Mathematics* (1928; rpt. La Salle, Ill.: Open Court, 1951), pp. 354–356.

9. See note 8.

10. By the Flemish scientist Simon Stevin (or Stevinius, 1548–1620).

11. Quoted in David Eugene Smith, *A Source Book in Mathematics* (1929; rpt. New York: Dover, 1959), p. 150.

12. Some other aspects of Napier's logarithms are discussed in Appendix 1.

13. Actually Napier came close to discovering the number $1/e$, defined as

the limit of $(1 - 1/n)^n$ as $n \rightarrow \infty$. As we have seen, his definition of logarithms is equivalent to the equation $N = 10^7(1 - 10^{-7})^L$. If we divide both N and L by 10^7 (which merely amounts to rescaling our variables), the equation becomes $N^* = [(1 - 10^{-7})^{10^7}]^{L^*}$, where $N^* = N/10^7$ and $L^* = L/10^7$. Since $(1 - 10^{-7})^{10^7} = (1 - 1/10^7)^{10^7}$ is very close to $1/e$, Napier's logarithms are virtually logarithms to the base $1/e$. The often-made statement that Napier discovered this base (or even e itself) is erroneous, however. As we have seen, he did not think in terms of a base, a concept that developed only later with the introduction of "common" (base 10) logarithms.

2

Recognition

The miraculous powers of modern calculation are due to three inventions: the Arabic Notation, Decimal Fractions, and Logarithms.
—FLORIAN CAJORI, *A History of Mathematics* (1893)

Napier published his invention in 1614 in a Latin treatise entitled *Mirifici logarithmorum canonis descriptio* (Description of the wonderful canon of logarithms). A later work, *Mirifici logarithmorum canonis constructio* (Construction of the wonderful canon of logarithms), was published posthumously by his son Robert in 1619. Rarely in the history of science has a new idea been received more enthusiastically. Universal praise was bestowed upon its inventor, and his invention was quickly adopted by scientists all across Europe and even in faraway China. One of the first to avail himself of logarithms was the astronomer Johannes Kepler, who used them with great success in his elaborate calculations of the planetary orbits.

Henry Briggs (1561–1631) was professor of geometry at Gresham College in London when word of Napier's tables reached him. So impressed was he by the new invention that he resolved to go to Scotland and meet the great inventor in person. We have a colorful account of their meeting by an astrologer named William Lilly (1602–1681):

One John Marr, an excellent mathematician and geometrician, had gone into Scotland before Mr. Briggs, purposely to be there when these two so learned persons should meet. Mr. Briggs appoints a certain day when to meet in Edinburgh; but failing thereof, the lord Napier was doubtful he would come. "Ah, John," said Napier, "Mr. Briggs will not now come." At that very moment one knocks at the gate; John Marr hastens down, and it proved Mr. Briggs to his great contentment. He brings Mr. Briggs up into the lord's chamber, where almost one quarter of an hour was spent, each beholding other with admiration, before one word was spoke. At last Briggs said: "My lord, I have undertaken this long journey purposely to see your person, and to know by what engine of wit or ingenuity you came first to think of this most

FIG. 1. Title page of the 1619 edition of Napier's *Mirifici logarithmorum canonis descriptio*, which also contains his *Constructio*.

excellent help in astronomy, viz. the logarithms; but, my lord, being by you found out, I wonder nobody found it out before, when now known it is so easy.[1]

At that meeting, Briggs proposed two modifications that would make Napier's tables more convenient: to have the logarithm of 1, rather than of 10^7, equal to 0; and to have the logarithm of 10 equal an appropriate power of 10. After considering several possibilities, they finally decided on log $10 = 1 = 10^0$. In modern phrasing this amounts to saying that if a positive number N is written as $N = 10^L$, then L is the Briggsian or "common" logarithm of N, written $\log_{10}N$ or simply log N. Thus was born the concept of *base*.[2]

Napier readily agreed to these suggestions, but by then he was already advanced in years and lacked the energy to compute a new set of tables. Briggs undertook this task, publishing his results in 1624 under the title *Arithmetica logarithmica*. His tables gave the logarithms to base 10 of all integers from 1 to 20,000 and from 90,000 to 100,000 to an accuracy of fourteen decimal places. The gap from 20,000 to 90,000 was later filled by Adriaan Vlacq (1600–1667), a Dutch publisher, and his additions were included in the second edition of the *Arithmetica logarithmica* (1628). With minor revisions, this work remained the basis for all subsequent logarithmic tables up to our century. Not until 1924 did work on a new set of tables, accurate to twenty places, begin in England as part of the tercentenary celebrations of the invention of logarithms. This work was completed in 1949.

Napier made other contributions to mathematics as well. He invented the rods or "bones" named after him—a mechanical device for performing multiplication and division—and devised a set of rules known as the "Napier analogies" for use in spherical trigonometry. And he advocated the use of the decimal point to separate the whole part of a number from its fractional part, a notation that greatly simplified the writing of decimal fractions. None of these accomplishments, however, compares in significance to his invention of logarithms. At the celebrations commemorating the three-hundredth anniversary of the occasion, held in Edinburgh in 1914, Lord Moulton paid him tribute: "The invention of logarithms came on the world as a bolt from the blue. No previous work had led up to it, foreshadowed it or heralded its arrival. It stands isolated, breaking in upon human thought abruptly without borrowing from the work of other intellects or following known lines of mathematical thought."[3] Napier died at his estate on 3 April 1617 at the age of sixty-seven and was buried at the church of St. Cuthbert in Edinburgh.[4]

Henry Briggs moved on to become, in 1619, the first Savilian Professor of Geometry at Oxford University, inaugurating a line of distinguished British scientists who would hold this chair, among them John Wallis, Edmond Halley, and Christopher Wren. At the same time, he kept his earlier position at Gresham College, occupying the chair that had been founded in 1596 by Sir Thomas Gresham, the earliest professorship of mathematics in England. He held both positions until his death in 1631.

One other person made claim to the title of inventor of logarithms. Jobst or Joost Bürgi (1552–1632), a Swiss watchmaker, constructed a table of logarithms on the same general scheme as Napier's, but with one significant difference: whereas Napier had used the common ratio $1 - 10^{-7}$, which is slightly less than 1, Bürgi used $1 + 10^{-4}$, a number slightly greater than 1. Hence Bürgi's logarithms *increase* with increasing numbers, while Napier's decrease. Like Napier, Bürgi was overly concerned with avoiding decimal fractions, making his definition of logarithms more complicated than necessary. If a positive integer N is written as $N = 10^8(1 + 10^{-4})^L$, then Bürgi called the number $10L$ (rather than L) the "red number" corresponding to the "black number" N. (In his table these numbers were actually printed in red and black, hence the nomenclature.) He placed the red numbers—that is, the logarithms—in the margin and the black numbers in the body of the page, in essence constructing a table of "antilogarithms." There is evidence that Bürgi arrived at his invention as early as 1588, six years before Napier began work on the same idea, but for some reason he did not publish it until 1620, when his table was issued anonymously in Prague. In academic matters the iron rule is "publish or perish." By delaying publication, Bürgi lost his claim for priority in a historic discovery. Today his name, except among historians of science, is almost forgotten.[5]

The use of logarithms quickly spread throughout Europe. Napier's *Descriptio* was translated into English by Edward Wright (ca. 1560–1615, an English mathematician and instrument maker) and appeared in London in 1616. Briggs's and Vlacq's tables of common logarithms were published in Holland in 1628. The mathematician Bonaventura Cavalieri (1598–1647), a contemporary of Galileo and one of the forerunners of the calculus, promoted the use of logarithms in Italy, as did Johannes Kepler in Germany. Interestingly enough, the next country to embrace the new invention was China, where in 1653 there appeared a treatise on logarithms by Xue Fengzuo, a disciple of the Polish Jesuit John Nicholas Smoguleçki (1611–1656). Vlacq's tables were reprinted in Beijing in 1713 in the *Lü-Li Yuan Yuan* (Ocean of calendar calculations). A later work, *Shu Li Ching Yün* (Collected basic principles of mathematics), was published in Beijing in 1722 and eventually reached Japan. All of this activity was a result of the Jesuits' presence in China and their commitment to the spread of Western science.[6]

No sooner had the scientific community adopted logarithms than some innovators realized that a mechanical device could be constructed to perform calculations with them. The idea was to use a ruler on which numbers are spaced in proportion to their logarithms. The first, rather primitive such device was built by Edmund Gunter (1581–1626), an English minister who later became professor of astronomy at Gresham College. His device appeared in 1620 and consisted of a single logarithmic scale along which distances could be measured and then added or subtracted with a pair of dividers. The idea of using *two* logarithmic scales that can be moved along each other originated with William Oughtred (1574–1660), who, like Gunter, was both a clergyman and a mathematician. Oughtred seems to have invented his device as early as 1622, but a description was not published until ten years later. In fact, Oughtred constructed two versions: a linear slide rule and a circular one, where the two scales were marked on discs that could rotate about a common pivot.[7]

Though Oughtred held no official university position, his contributions to mathematics were substantial. In his most influential work, the *Clavis mathematicae* (1631), a book on arithmetic and algebra, he introduced many new mathematical symbols, some of which are still in use today. (Among them is the symbol × for multiplication, to which Leibniz later objected because of its similarity to the letter x; two other symbols that can still be seen occasionally are : : to denote a proportion and ∼ for "the difference between.") Today we take for granted the numerous symbols that appear in the mathematical literature, but each has a history of its own, often reflecting the state of mathematics at the time. Symbols were sometimes invented at the whim of a mathematician; but more often they were the result of a slow evolution, and Oughtred was a major player in this process. Another mathematician who did much to improve mathematical notation was Leonhard Euler, who will figure prominently later in our story.

About Oughtred's life there are many stories. As a student at King's College in Cambridge he spent day and night on his studies, as we know from his own account: "The time which over and above those usuall studies I employed upon the Mathematicall sciences, I redeemed night by night from my naturall sleep, defrauding my body, and inuring it to watching, cold, and labour, while most others tooke their rest."[8] We also have the colorful account of Oughtred in John Aubrey's entertaining (though not always reliable) *Brief Lives*:

He was a little man, had black haire, and blacke eies (with a great deal of spirit). His head was always working. He would drawe lines and diagrams on the dust . . . did use to lye a bed till eleaven or twelve a clock. . . . Studyed late at night; went not to bed till ll a clock; had his tinder box by him; and on the top of his bed-staffe, he had his inke-horne fix't. He slept but little. Sometimes he went not to bed in two or three nights.[9]

Though he seems to have violated every principle of good health, Oughtred died at the age of eighty-six, reportedly of joy upon hearing that King Charles II had been restored to the throne.

As with logarithms, claims of priority for inventing the slide rule did not go unchallenged. In 1630 Richard Delamain, a student of Oughtred, published a short work, *Grammelogia, or The Mathemati-*

call Ring, in which he described a circular slide rule he had invented. In the preface, addressed to King Charles I (to whom he sent a slide rule and a copy of the book), Delamain mentions the ease of operation of his device, noting that it was "fit for use . . . as well on Horse backe as on Foot."[10] He duly patented his invention, believing that his copyright and his name in history would thereby be secured. However, another pupil of Oughtred, William Forster, claimed that he had seen Oughtred's slide rule at Delamain's home some years earlier, implying that Delamain had stolen the idea from Oughtred. The ensuing series of charges and countercharges was to be expected, for nothing can be more damaging to a scientist's reputation than an accusation of plagiarism. It is now accepted that Oughtred was indeed the inventor of the slide rule, but there is no evidence to support Forster's claim that Delamain stole the invention. In any event, the dispute has long since been forgotten, for it was soon overshadowed by a far more acrimonious dispute over an invention of far greater importance: the calculus.

The slide rule, in its many variants, would be the faithful companion of every scientist and engineer for the next 350 years, proudly given by parents to their sons and daughters upon graduation from college. Then in the early 1970s the first electronic hand-held calculators appeared on the market, and within ten years the slide rule was obsolete. (In 1980 a leading American manufacturer of scientific instruments, Keuffel & Esser, ceased production of its slide rules, for which it had been famous since 1891.[11]) As for logarithmic tables, they have fared a little better: one can still find them at the back of algebra textbooks, a mute reminder of a tool that has outlived its usefulness. It won't be long, however, before they too will be a thing of the past.

But if logarithms have lost their role as the centerpiece of computational mathematics, the logarithmic *function* remains central to almost every branch of mathematics, pure or applied. It shows up in a host of applications, ranging from physics and chemistry to biology, psychology, art, and music. Indeed, one contemporary artist, M. C. Escher, has made the the logarithmic function—disguised as a spiral—a central theme of much of his work (see p. 138).

In the second edition of Edward Wright's translation of Napier's *Descriptio* (London, 1618), in an appendix probably written by Oughtred, there appears the equivalent of the statement that $\log_e 10 = 2.302585$.[12] This seems to be the first explicit recognition of the role of the number e in mathematics. But where did this number come from? Wherein lies its importance? To answer these questions, we must now turn to a subject that at first seems far removed from exponents and logarithms: the mathematics of finance.

NOTES AND SOURCES

1. Quoted in Eric Temple Bell, *Men of Mathematics* (1937; rpt. Harmondsworth: Penguin Books, 1965), 2:580; Edward Kasner and James Newman, *Mathematics and the Imagination* (New York: Simon and Schuster, 1958), p. 81. The original appears in Lilly's *Description of his Life and Times* (1715).

2. See George A. Gibson, "Napier's Logarithms and the Change to Briggs's Logarithms," in *Napier Tercentenary Memorial Volume*, ed. Cargill Gilston Knott (London: Longmans, Green and Company, 1915), p. 111. See also Julian Lowell Coolidge, *The Mathematics of Great Amateurs* (New York: Dover, 1963), ch. 6, esp. pp. 77–79.

3. Inaugural address, "The Invention of Logarithms," in *Napier Tercentenary Memorial Volume*, p. 3.

4. *Handbook of the Napier Tercentenary Celebration, or Modern Instruments and Methods of Calculation*, ed. E. M. Horsburgh (1914; Los Angeles: Tomash Publishers, 1982), p. 16. Section A is a detailed account of Napier's life and work.

5. On the question of priority, see Florian Cajori, "Algebra in Napier's Day and Alleged Prior Inventions of Logarithms," in *Napier Tercentenary Memorial Volume*, p. 93.

6. Joseph Needham, *Science and Civilisation in China* (Cambridge: Cambridge University Press, 1959), 3:52–53.

7. David Eugene Smith, *A Source Book in Mathematics* (1929; rpt. New York: Dover, 1959), pp. 160–164.

8. Quoted in David Eugene Smith, *History of Mathematics*, 2 vols. (1923; New York: Dover, 1958), 1:393.

9. John Aubrey, *Brief Lives*, 2:106 (as quoted by Smith, *History of Mathematics*, 1:393).

10. Quoted in Smith, *A Source Book in Mathematics*, pp. 156–159.

11. *New York Times*, 3 January 1982.

12. Florian Cajori, *A History of Mathematics* (1893), 2d ed. (New York: Macmillan, 1919), p. 153; Smith, *History of Mathematics*, 2:517.

Computing with Logarithms

For many of us—at least those who completed our college education after 1980—logarithms are a theoretical subject, taught in an introductory algebra course as part of the function concept. But until the late 1970s logarithms were still widely used as a computational device, virtually unchanged from Briggs's common logarithms of 1624. The advent of the hand-held calculator has made their use obsolete.

Let us say it is the year 1970 and we are asked to compute the expression

$$x = \sqrt[3]{(493.8 \cdot 23.67^2/5.104)}.$$

For this task we need a table of four-place common logarithms (which can still be found at the back of most algebra textbooks). We also need to use the laws of logarithms:

$$\log (ab) = \log a + \log b, \log (a/b) = \log a - \log b,$$
$$\log a^n = n \log a,$$

where a and b denote any positive numbers and n any real number; here "log" stands for common logarithm—that is, logarithm base 10—although any other base for which tables are available could be used.

Before we start the computation, let us recall the definition of logarithm: If a positive number N is written as $N = 10^L$, then L is the logarithm (base 10) of N, written log N. Thus the equations $N = 10^L$ and $L = \log N$ are equivalent—they give exactly the same information. Since $1 = 10^0$ and $10 = 10^1$, we have log $1 = 0$ and log $10 = 1$. Therefore, the logarithm of any number between 1 (inclusive) and 10 (exclusive) is a positive fraction, that is, a number of the form $0 . a\,b\,c \ldots$; in the same way, the logarithm of any number between 10 (inclusive) and 100 (exclusive) is of the form $1 . a\,b\,c \ldots$, and so on. We summarize this as:

Range of N	log N
$1 \le N < 10,$	$0 . a\,b\,c \ldots$
$10 \le N < 100,$	$1 . a\,b\,c \ldots$
$100 \le N < 1,000,$	$2 . a\,b\,c \ldots$
$\ldots$	

(The table can be extended backward to include fractions, but we have not done so here in order to keep the discussion simple.) Thus, if a logarithm is written as $\log N = p \cdot abc \ldots$, the integer p tells us in what range of powers of 10 the number N lies; for example, if we are told that $\log N = 3.456$, we can conclude that N lies between 1,000 and 10,000. The actual value of N is determined by the fractional part $. abc \ldots$ of the logarithm. The integral part p of $\log N$ is called its *characteristic* , and the fractional part $. abc \ldots$ its *mantissa*.[1] A table of logarithms usually gives only the mantissa; it is up to the user to determine the characteristic. Note that two logarithms with the same mantissa but different characteristics correspond to two numbers having the same digits but a different position of the decimal point. For example, $\log N = 0.267$ corresponds to $N = 1.849$, whereas $\log N = 1.267$ corresponds to $N = 18.49$. This becomes clear if we write these two statements in exponential form: $10^{0.267} = 1.849$, while $10^{1.267} = 10 \cdot 10^{0.267} = 10 \cdot 1.849 = 18.49$.

We are now ready to start our computation. We begin by writing x in a form more suitable for logarithmic computation by replacing the radical with a fractional exponent:

$$x = (493.8 \cdot 23.67^2 / 5.104)^{1/3}.$$

Taking the logarithm of both sides, we have

$$\log x = (1/3)[\log 493.8 + 2 \log 23.67 - \log 5.104].$$

We now find each logarithm, using the Proportional Parts section of the table to add the value given there to that given in the main table. Thus, to find $\log 493.8$ we locate the row that starts with 49, move across to the column headed by 3 (where we find 6928), and then look under the column 8 in the Proportional Parts to find the entry 7. We add this entry to 6928 and get 6935. Since 493.8 is between 100 and 1,000, the characteristic is 2; we thus have $\log 493.8 = 2.6935$. We do the same for the other numbers. It is convenient to do the computation in a table:

	N	$\log N$
	23.67 $\rightarrow$	1.3742
		$\times$ 2
		2.7484
	493.8 $\rightarrow$	$+$ 2.6935
		5.4419
	5.104 $\rightarrow$	$-$ 0.7079
		4.7340 : 3
Answer:	37.84 $\leftarrow$	1.5780

N	0	1	2	3	4	5	6	7	8	9	Proportional Parts								
											1	2	3	4	5	6	7	8	9
10	0000	0043	0086	0128	0170	0212	0253	0294	0334	0374	4	8	12	17	21	25	29	33	37
11	0414	0453	0492	0531	0569	0607	0645	0682	0719	0755	4	8	11	15	19	23	26	30	34
12	0792	0828	0864	0899	0934	0969	1004	1038	1072	1106	3	7	10	14	17	21	24	28	31
13	1139	1173	1206	1239	1271	1303	1335	1367	1399	1430	3	6	10	13	16	19	23	26	29
14	1461	1492	1523	1553	1584	1614	1644	1673	1703	1732	3	6	9	12	15	18	21	24	27
15	1761	1790	1818	1847	1875	1903	1931	1959	1987	2014	3	6	8	11	14	17	20	22	25
16	2041	2068	2095	2122	2148	2175	2201	2227	2253	2279	3	5	8	11	13	16	18	21	24
17	2304	2330	2355	2380	2405	2430	2455	2480	2504	2529	2	5	7	10	12	15	17	20	22
18	2553	2577	2601	2625	2648	2672	2695	2718	2742	2765	2	5	7	9	12	14	16	19	21
19	2788	2810	2833	2856	2878	2900	2923	2945	2967	2989	2	4	7	9	11	13	16	18	20
20	3010	3032	3054	3075	3096	3118	3139	3160	3181	3201	2	4	6	8	11	13	15	17	19
21	3222	3243	3263	3284	3304	3324	3345	3365	3385	3404	2	4	6	8	10	12	14	16	18
22	3424	3444	3464	3483	3502	3522	3541	3560	3579	3598	2	4	6	8	10	12	14	15	17
23	3617	3636	3655	3674	3692	3711	3729	3747	3766	3784	2	4	6	7	9	11	13	15	17
24	3802	3820	3838	3856	3874	3892	3909	3927	3945	3962	2	4	5	7	9	11	12	14	16
25	3979	3997	4014	4031	4048	4065	4082	4099	4116	4133	2	3	5	7	9	10	12	14	15
26	4150	4166	4183	4200	4216	4232	4249	4265	4281	4298	2	3	5	7	8	10	11	13	15
27	4314	4330	4346	4362	4378	4393	4409	4425	4440	4456	2	3	5	6	8	9	11	13	14
28	4472	4487	4502	4518	4533	4548	4564	4579	4594	4609	2	3	5	6	8	9	11	12	14
29	4624	4639	4654	4669	4683	4698	4713	4728	4742	4757	1	3	4	6	7	9	10	12	13
30	4771	4786	4800	4814	4829	4843	4857	4871	4886	4900	1	3	4	6	7	9	10	11	13
31	4914	4928	4942	4955	4969	4983	4997	5011	5024	5038	1	3	4	6	7	8	10	11	12
32	5051	5065	5079	5092	5105	5119	5132	5145	5159	5172	1	3	4	5	7	8	9	11	12
33	5185	5198	5211	5224	5237	5250	5263	5276	5289	5302	1	3	4	5	6	8	9	10	12
34	5315	5328	5340	5353	5366	5378	5391	5403	5416	5428	1	3	4	5	6	8	9	10	11
35	5441	5453	5465	5478	5490	5502	5514	5527	5539	5551	1	2	4	5	6	7	9	10	11
36	5563	5575	5587	5599	5611	5623	5635	5647	5658	5670	1	2	4	5	6	7	8	10	11
37	5682	5694	5705	5717	5729	5740	5752	5763	5775	5786	1	2	3	5	6	7	8	9	10
38	5798	5809	5821	5832	5843	5855	5866	5877	5888	5899	1	2	3	5	6	7	8	9	10
39	5911	5922	5933	5944	5955	5966	5977	5988	5999	6010	1	2	3	4	5	7	8	9	10
40	6021	6031	6042	6053	6064	6075	6085	6096	6107	6117	1	2	3	4	5	6	8	9	10
41	6128	6138	6149	6160	6170	6180	6191	6201	6212	6222	1	2	3	4	5	6	7	8	9
42	6232	6243	6253	6263	6274	6284	6294	6304	6314	6325	1	2	3	4	5	6	7	8	9
43	6335	6345	6355	6365	6375	6385	6395	6405	6415	6425	1	2	3	4	5	6	7	8	9
44	6435	6444	6454	6464	6474	6484	6493	6503	6513	6522	1	2	3	4	5	6	7	8	9
45	6532	6542	6551	6561	6571	6580	6590	6599	6609	6618	1	2	3	4	5	6	7	8	9
46	6628	6637	6646	6656	6665	6675	6684	6693	6702	6712	1	2	3	4	5	6	7	7	8
47	6721	6730	6739	6749	6758	6767	6776	6785	6794	6803	1	2	3	4	5	5	6	7	8
48	6812	6821	6830	6839	6848	6857	6866	6875	6884	6893	1	2	3	4	4	5	6	7	8
49	6902	6911	6920	6928	6937	6946	6955	6964	6972	6981	1	2	3	4	4	5	6	7	8
50	6990	6998	7007	7016	7024	7033	7042	7050	7059	7067	1	2	3	3	4	5	6	7	8
51	7076	7084	7093	7101	7110	7118	7126	7135	7143	7152	1	2	3	3	4	5	6	7	8
52	7160	7168	7177	7185	7193	7202	7210	7218	7226	7235	1	2	2	3	4	5	6	7	7
53	7243	7251	7259	7267	7275	7284	7292	7300	7308	7316	1	2	2	3	4	5	6	6	7
54	7324	7332	7340	7348	7356	7364	7372	7380	7388	7396	1	2	2	3	4	5	6	6	7
N	0	1	2	3	4	5	6	7	8	9	1	2	3	4	5	6	7	8	9

Four-Place Logarithms

For the last step we used a table of *antilogarithms*—logarithms in reverse. We look up the number .5780 (the mantissa) and find the entry 3784; since the characteristic of 1.5780 is 1, we know that the number must be between 10 and 100. Thus x = 37.84, rounded to two places.

p	0	1	2	3	4	5	6	7	8	9	Proportional Parts 1	2	3	4	5	6	7	8	9
.50	3162	3170	3177	3184	3192	3199	3206	3214	3221	3228	1	1	2	3	4	4	5	6	7
.51	3236	3243	3251	3258	3266	3273	3281	3289	3296	3304	1	2	2	3	4	5	5	6	7
.52	3311	3319	3327	3334	3342	3350	3357	3365	3373	3381	1	2	2	3	4	5	5	6	7
.53	3388	3396	3404	3412	3420	3428	3436	3443	3451	3459	1	2	2	3	4	5	6	6	7
.54	3467	3475	3483	3491	3499	3508	3516	3524	3532	3540	1	2	2	3	4	5	6	6	7
.55	3548	3556	3565	3573	3581	3589	3597	3606	3614	3622	1	2	2	3	4	5	6	7	7
.56	3631	3639	3648	3656	3664	3673	3681	3690	3698	3707	1	2	3	3	4	5	6	7	8
.57	3715	3724	3733	3741	3750	3758	3767	3776	3784	3793	1	2	3	3	4	5	6	7	8
.58	3802	3811	3819	3828	3837	3846	3855	3864	3873	3882	1	2	3	4	4	5	6	7	8
.59	3890	3899	3908	3917	3926	3936	3945	3954	3963	3972	1	2	3	4	5	5	6	7	8
.60	3981	3990	3999	4009	4018	4027	4036	4046	4055	4064	1	2	3	4	5	6	6	7	8
.61	4074	4083	4093	4102	4111	4121	4130	4140	4150	4159	1	2	3	4	5	6	7	8	9
.62	4169	4178	4188	4198	4207	4217	4227	4236	4246	4256	1	2	3	4	5	6	7	8	9
.63	4266	4276	4285	4295	4305	4315	4325	4335	4345	4355	1	2	3	4	5	6	7	8	9
.64	4365	4375	4385	4395	4406	4416	4426	4436	4446	4457	1	2	3	4	5	6	7	8	9
.65	4467	4477	4487	4498	4508	4519	4529	4539	4550	4560	1	2	3	4	5	6	7	8	9
.66	4571	4581	4592	4603	4613	4624	4634	4645	4656	4667	1	2	3	4	5	6	7	9	10
.67	4677	4688	4699	4710	4721	4732	4742	4753	4764	4775	1	2	3	4	5	7	8	9	10
.68	4786	4797	4808	4819	4831	4842	4853	4864	4875	4887	1	2	3	4	6	7	8	9	10
.69	4898	4909	4920	4932	4943	4955	4966	4977	4989	5000	1	2	3	5	6	7	8	9	10
.70	5012	5023	5035	5047	5058	5070	5082	5093	5105	5117	1	2	4	5	6	7	8	9	11
.71	5129	5140	5152	5164	5176	5188	5200	5212	5224	5236	1	2	4	5	6	7	8	10	11
.72	5248	5260	5272	5284	5297	5309	5321	5333	5346	5358	1	2	4	5	6	7	9	10	11
.73	5370	5383	5395	5408	5420	5433	5445	5458	5470	5483	1	3	4	5	6	8	9	10	11
.74	5495	5508	5521	5534	5546	5559	5572	5585	5598	5610	1	3	4	5	6	8	9	10	12
.75	5623	5636	5649	5662	5675	5689	5702	5715	5728	5741	1	3	4	5	7	8	9	10	12
.76	5754	5768	5781	5794	5808	5821	5834	5848	5861	5875	1	3	4	5	7	8	9	11	12
.77	5888	5902	5916	5929	5943	5957	5970	5984	5998	6012	1	3	4	5	7	8	10	11	12
.78	6026	6039	6053	6067	6081	6095	6109	6124	6138	6152	1	3	4	6	7	8	10	11	13
.79	6166	6180	6194	6209	6223	6237	6252	6266	6281	6295	1	3	4	6	7	9	10	11	13
.80	6310	6324	6339	6353	6368	6383	6397	6412	6427	6442	1	3	4	6	7	9	10	12	13
.81	6457	6471	6486	6501	6516	6531	6546	6561	6577	6592	2	3	5	6	8	9	11	12	14
.82	6607	6622	6637	6653	6668	6683	6699	6714	6730	6745	2	3	5	6	8	9	11	12	14
.83	6761	6776	6792	6808	6823	6839	6855	6871	6887	6902	2	3	5	6	8	9	11	13	14
.84	6918	6934	6950	6966	6982	6998	7015	7031	7047	7063	2	3	5	6	8	10	11	13	15
.85	7079	7096	7112	7129	7145	7161	7178	7194	7211	7228	2	3	5	7	8	10	12	13	15
.86	7244	7261	7278	7295	7311	7328	7345	7362	7379	7396	2	3	5	7	8	10	12	13	15
.87	7413	7430	7447	7464	7482	7499	7516	7534	7551	7568	2	3	5	7	9	10	12	14	16
.88	7586	7603	7621	7638	7656	7674	7691	7709	7727	7745	2	4	5	7	9	11	12	14	16
.89	7762	7780	7798	7816	7834	7852	7870	7889	7907	7925	2	4	5	7	9	11	13	14	16
.90	7943	7962	7980	7998	8017	8035	8054	8072	8091	8110	2	4	6	7	9	11	13	15	17
.91	8128	8147	8166	8185	8204	8222	8241	8260	8279	8299	2	4	6	8	9	11	13	15	17
.92	8318	8337	8356	8375	8395	8414	8433	8453	8472	8492	2	4	6	8	10	12	14	15	17
.93	8511	8531	8551	8570	8590	8610	8630	8650	8670	8690	2	4	6	8	10	12	14	16	18
.94	8710	8730	8750	8770	8790	8810	8831	8851	8872	8892	2	4	6	8	10	12	14	16	18
.95	8913	8933	8954	8974	8995	9016	9036	9057	9078	9099	2	4	6	8	10	12	15	17	19
.96	9120	9141	9162	9183	9204	9226	9247	9268	9290	9311	2	4	6	8	11	13	15	17	19
.97	9333	9354	9376	9397	9419	9441	9462	9484	9506	9528	2	4	7	9	11	13	15	17	20
.98	9550	9572	9594	9616	9638	9661	9683	9705	9727	9750	2	4	7	9	11	13	16	18	20
.99	9772	9795	9817	9840	9863	9886	9908	9931	9954	9977	2	5	7	9	11	14	16	18	20
p	0	1	2	3	4	5	6	7	8	9	1	2	3	4	5	6	7	8	9

Four-Place Antilogarithms

Sounds complicated? Yes, if you have been spoiled by the calculator. With some experience, the above calculation can be completed in two or three minutes; on a calculator it should take no more than a few seconds (and you get the answer correct to six places, 37.845331). But let us not forget that from 1614, the year logarithms were invented, to around 1945, when the first electronic computers became operative, logarithms—or their mechanical equivalent, the slide rule—were practically the only way to perform such calculations. No wonder the scientific community embraced them with such enthusiasm. As the eminent mathematician Pierre Simon Laplace said, "By shortening the labors, the invention of logarithms doubled the life of the astronomer."

NOTE

1. The terms *characteristic* and *mantissa* were suggested by Henry Briggs in 1624. The word *mantissa* is a late Latin term of Etruscan origin, meaning a makeweight, a small weight added to a scale to bring the weight to a desired value. See David Eugene Smith, *History of Mathematics*, 2 vols. (1923; rpt. New York: Dover, 1958), 2:514.

3

Financial Matters

If thou lend money to any of My people, . . .

thou shalt not be to him as a creditor;

neither shall ye lay upon him interest.

—Exodus 22:24

From time immemorial money matters have been at the center of human concerns. No other aspect of life has a more mundane character than the urge to acquire wealth and achieve financial security. So it must have been with some surprise that an anonymous mathematician—or perhaps a merchant or moneylender—in the early seventeenth century noticed a curious connection between the way money grows and the behavior of a certain mathematical expression at infinity.

Central to any consideration of money is the concept of *interest*, or money paid on a loan. The practice of charging a fee for borrowing money goes back to the dawn of recorded history; indeed, much of the earliest mathematical literature known to us deals with questions related to interest. For example, a clay tablet from Mesopotamia, dated to about 1700 B.C. and now in the Louvre, poses the following problem: How long will it take for a sum of money to double if invested at 20 percent interest rate compounded annually?[1] To formulate this problem in the language of algebra, we note that at the end of each year the sum grows by 20 percent, that is, by a factor of 1.2; hence after x years the sum will grow by a factor of 1.2^x. Since this is to be equal to twice the original sum, we have $1.2^x = 2$ (note that the original sum does not enter the equation).

Now to solve this equation—that is, to remove x from the exponent—we must use logarithms, which the Babylonians did not have. Nevertheless, they were able to find an approximate solution by observing that $1.2^3 = 1.728$, while $1.2^4 = 2.0736$; so x must have a value between 3 and 4. To narrow this interval, they used a process known as linear interpolation—finding a number that divides the interval from 3 to 4 in the same ratio as 2 divides the interval from 1.728 to

2.0736. This leads to a linear (first-degree) equation in x, which can easily be solved using elementary algebra. But the Babylonians did not possess our modern algebraic techniques, and to find the required value was no simple task for them. Still, their answer, $x = 3.7870$, comes remarkably close to the correct value, 3.8018 (that is, about three years, nine months, and eighteen days). We should note that the Babylonians did not use our decimal system, which came into use only in the early Middle Ages; they used the *sexagesimal* system, a numeration system based on the number 60. The answer on the Louvre tablet is given as 3;47,13,20, which in the sexagesimal system means $3 + 47/60 + 13/60^2 + 20/60^3$, or very nearly 3.7870.[2]

In a way, the Babylonians did use a logarithmic table of sorts. Among the surviving clay tablets, some list the first ten powers of the numbers 1/36, 1/16, 9, and 16 (the first two expressed in the sexagesimal system as 0;1,40 and 0;3,45)—all perfect squares. Inasmuch as such a table lists the powers of a number rather than the exponent, it is really a table of antilogarithms, except that the Babylonians did not use a single, standard base for their powers. It seems that these tables were compiled to deal with a specific problem involving compound interest rather than for general use.[3]

Let us briefly examine how compound interest works. Suppose we invest $100 (the "principal") in an account that pays 5 percent interest, compounded annually. At the end of one year, our balance will be $100 \times 1.05 = \$105$. The bank will then consider this new amount as a new principal that has just been reinvested at the same rate. At the end of the second year the balance will therefore be $105 \times 1.05 = \$110.25$, at the end of the third year $110.25 \times 1.05 = \$115.76$, and so on. (Thus, not only the principal bears annual interest but also the *interest* on the principal—hence the phrase "compound interest.") We see that our balance grows in a geometric progression with the common ratio 1.05. By contrast, in an account that pays *simple* interest the annual rate is applied to the *original* principal and is therefore the same every year. Had we invested our $100 at 5 percent simple interest, our balance would increase each year by $5, giving us the arithmetic progression 100, 105, 110, 115, and so on. Clearly, money invested at compound interest—regardless of the rate—will eventually grow faster than if invested at simple interest.

From this example it is easy to see what happens in the general case. Suppose we invest a principal of P dollars in an account that pays r percent interest rate compounded annually (in the computations we always express r as a decimal, for example, 0.05 instead of 5 percent). This means that at the end of the first year our balance will be $P(1 + r)$, at the end of the second year, $P(1 + r)^2$, and so on until after t years the balance will be $P(1 + r)^t$. Denoting this amount by S, we arrive at the formula

$$S = P(1 + r)^t. \tag{1}$$

This formula is the basis of virtually all financial calculations, whether they apply to bank accounts, loans, mortgages, or annuities.

Some banks compute the accrued interest not once but several times a year. If, for example, an annual interest rate of 5 percent is compounded semiannually, the bank will use one-half of the annual interest rate as the rate *per period*. Hence, in one year a principal of $100 will be compounded twice, each time at the rate of 2.5 percent; this will amount to 100×1.025^2 or $105.0625, about six cents more than the same principal would yield if compounded annually at 5 percent.

In the banking industry one finds all kinds of compounding schemes—annual, semiannual, quarterly, weekly, and even daily. Suppose the compounding is done n times a year. For each "conversion period" the bank uses the annual interest rate *divided by n*, that is, r/n. Since in t years there are (nt) conversion periods, a principal P will after t years yield the amount

$$S = P(1 + r/n)^{nt}. \tag{2}$$

Of course, equation 1 is just a special case of equation 2—the case where n = 1.

It would be interesting to compare the amounts of money a given principal will yield after one year for different conversion periods, assuming the same annual interest rate. Let us take as an example $P = 100 and $r = 5$ percent $= 0.05$. Here a hand-held calculator will be useful. If the calculator has an exponentiation key (usually denoted by y^x), we can use it to compute the desired values directly; otherwise we will have to use repeated multiplication by the factor $(1 + 0.05/n)$. The results, shown in table 3.1, are quite surprising. As we see, a principal of $100 compounded daily yields just thirteen cents more than when compounded annually, and about *one cent*

TABLE 3.1. $100 Invested for One Year at 5 Percent Annual Interest Rate at Different Conversion Periods

Conversion Period	n	r/n	S
Annually	1	0.05	$105.00
Semiannually	2	0.025	$105.06
Quarterly	4	0.0125	$105.09
Monthly	12	0.004166	$105.12
Weekly	52	0.0009615	$105.12
Daily	365	0.0001370	$105.13

more than when compounded monthly or weekly! It hardly makes a difference in which account we invest our money.[4]

To explore this question further, let us consider a special case of equation 2, the case when $r = 1$. This means an annual interest rate of 100 percent, and certainly no bank has ever come up with such a generous offer. What we have in mind, however, is not an actual situation but a hypothetical case, one that has far-reaching mathematical consequences. To simplify our discussion, let us assume that $P = \$1$ and $t = 1$ year. Equation 2 then becomes

$$S = (1 + 1/n)^n \tag{3}$$

and our aim is to investigate the behavior of this formula for increasing values of n. The results are given in table 3.2.

TABLE 3.2

n	$(1 + 1/n)^n$
1	2
2	2.25
3	2.37037
4	2.44141
5	2.48832
10	2.59374
50	2.69159
100	2.70481
1,000	2.71692
10,000	2.71815
100,000	2.71827
1,000,000	2.71828
10,000,000	2.71828

It looks as if any further increase in n will hardly affect the outcome—the changes will occur in less and less significant digits.

But will this pattern go on? Is it possible that no matter how large n is, the values of $(1 + 1/n)^n$ will settle somewhere around the number 2.71828? This intriguing possibility is indeed confirmed by careful mathematical analysis (see Appendix 2). We do not know who first noticed the peculiar behavior of the expression $(1 + 1/n)^n$ as n tends to infinity, so the exact date of birth of the number that would later be denoted by e remains obscure. It seems likely, however, that its origins go back to the early seventeenth century, around the time when Napier invented his logarithms. (As we have seen, the second edition of Edward Wright's translation of Napier's *Descriptio* [1618] contained an indirect reference to e.) This period was marked by enormous growth in international trade, and financial transactions of all sorts proliferated; as a result, a great deal of attention was paid to the law of compound interest, and it is possible that the number *e* received its first recognition in this context. We shall soon see, however, that questions unrelated to compound interest also led to the same number at about the same time. But before we turn to these questions, we would do well to take a closer look at the mathematical process that is at the root of *e*: the limit process.

NOTES AND SOURCES

1. Howard Eves, *An Introduction to the History of Mathematics* (1964; rpt. Philadelphia: Saunders College Publishing, 1983), p. 36.

2. Carl B. Boyer, *A History of Mathematics*, rev. ed. (New York: John Wiley, 1989), p. 36.

3. Ibid., p. 35.

4. Of course, the difference is still proportional to the principal. If we invest $1,000,000 instead of $100, our balance at the end of the first year will be $1,050,000 if compounded annually, compared to $1,051,267.50 if compounded daily—a difference of $1267.50. You are always better off to be rich!

4

To the Limit, If It Exists

I saw, as one might see the transit of Venus, a quantity
passing through infinity and changing its sign from plus to
minus. I saw exactly how it happened . . . but it was after
dinner and I let it go.
—SIR WINSTON CHURCHILL, *My Early Life* (1930)

At first thought, the peculiar behavior of the expression $(1 + 1/n)^n$ for large values of n must seem puzzling indeed. Suppose we consider only the expression inside the parentheses, $1 + 1/n$. As n increases, $1/n$ gets closer and closer to 0 and so $1 + 1/n$ gets closer and closer to 1, although it will always be greater than 1. Thus we might be tempted to conclude that for "really large" n (whatever "really large" means), $1 + 1/n$, to every purpose and extent, may be replaced by 1. Now 1 raised to any power is always equal to 1, so it seems that $(1 + 1/n)^n$ for large n should approach the number 1. Had this been the case, there would be nothing more for us to say about the subject.

But suppose we follow a different approach. We know that when a number greater than 1 is raised to increasing powers, the result becomes larger and larger. Since $1 + 1/n$ is always greater than 1, we might conclude that $(1 + 1/n)^n$, for large values of n, will grow without bound, that is, tend to infinity. Again, that would be the end of our story.

That this kind of reasoning is seriously flawed can already be seen from the fact that, depending on our approach, we arrived at two different results: 1 in the first case and infinity in the second. In mathematics, the end result of any *valid* numerical operation, regardless of how it was arrived at, must always be the same. For example, we can evaluate the expression $2 \cdot (3 + 4)$ either by first adding 3 and 4 to get 7 and then doubling the result, or by first doubling each of the numbers 3 and 4 and then adding the results. In either case we get 14. Why, then, did we get two different results for $(1 + 1/n)^n$?

The answer lies in the word *valid*. When we computed the expres-

Basic rule

sion $2 \cdot (3 + 4)$ by the second method, we tacitly used one of the fundamental laws of arithmetic, the distributive law, which says that for any three numbers x, y, and z the equation $x \cdot (y + z) = x \cdot y + x \cdot z$ is always true. To go from the left side of this equation to the right side is a valid operation. An example of an *invalid* operation is to write $\sqrt{9 + 16} = 3 + 4 = 7$, a mistake that beginning algebra students often make. The reason is that taking a square root is not a distributive operation; indeed, the only proper way of evaluating $\sqrt{9 + 16}$ is *first* to add the numbers under the radical sign and *then* take the square root: $\sqrt{9 + 16} = \sqrt{25} = 5$. Our handling of the expression $(1 + 1/n)^n$ was equally invalid, because we wrongly played with one of the most fundamental concepts of mathematical analysis: the concept of *limit*.

When we say that a sequence of numbers a_1, a_2, a_3, . . . , a_n, . . . tends to a limit L as n tends to infinity, we mean that as n grows larger and larger, the terms of the sequence get closer and closer to the number L. Put in different words, we can make the difference (in absolute value) between a_n and L as small as we please by going out far enough in our sequence—that is, by choosing n to be sufficiently large. Take, for example, the sequence 1, 1/2, 1/3, 1/4, . . . , whose general term is $a_n = 1/n$. As n increases, the terms get closer and closer to 0. This means that the difference between $1/n$ and the limit 0 (that is, just $1/n$) can be made as small as we please if we choose n large enough. Say that we want $1/n$ to be less than 1/1,000; all we need to do is make n *greater* than 1,000. If we want $1/n$ to be less than 1/1,000,000, we simply choose any n greater than 1,000,000. And so on. We express this situation by saying that $1/n$ tends to 0 as n increases without bound, and we write $1/n \to 0$ as n $\to \infty$. We also use the abbreviated notation

$$\lim_{n \to \infty} \frac{1}{n} = 0.$$

A word of caution is necessary, however: the expression $\lim_{n \to \infty} 1/n = 0$ says only that the *limit* of $1/n$ as $n \to \infty$ is 0; it does not say that $1/n$ itself will ever be equal to 0—in fact, it will not. This is the very essence of the limit concept: a sequence of numbers can *approach* a limit as closely as we please, but it will never actually reach it.[1]

For the sequence $1/n$, the outcome of the limiting process is quite predictable. In many cases, however, it may not be immediately clear what the limiting value will be or whether there is a limit at all. For example, the sequence $a_n = (2n + 1)/(3n + 4)$, whose terms for $n = 1$, 2, 3, . . . are 3/7, 5/10, 7/13, . . . , tends to the limit 2/3 as $n \to \infty$. This can be seen by dividing the numerator and denominator by n, giving us the equivalent expression $a_n = (2 + 1/n)/(3 + 4/n)$. As $n \to \infty$, both $1/n$ and $4/n$ tend to 0, so that the entire expression tends to 2/3. On the other hand, the sequence $a_n = (2n^2 + 1)/(3n + 4)$, whose members are 3/7, 9/10, 19/13, ... , grows without bound as $n \to \infty$. This is because the term n^2 causes the numerator to grow at a faster rate than the denominator. We express this fact by writing $\lim_{n \to \infty} a_n = \infty$, although strictly speaking the sequence does not have a limit. A limit—if it exists—must be a definite real number, and infinity is not a real number.

For centuries, mathematicians and philosophers have been intrigued by the concept of infinity. Is there a number greater than all numbers? If so, just how large is such a "number"? Can we calculate with it as we do with ordinary numbers? And on the small scale of things, can we divide a quantity—say a number or a line segment—again and again into smaller quantities, or will we eventually reach an indivisible part, a mathematical atom that cannot be further split? Questions such as these troubled the philosophers of ancient Greece more than two thousand years ago, and they still trouble us today—witness the never ending search for the elementary particles, those elusive building blocks from which all matter is believed to be constructed.

That we cannot use the symbol for infinity, ∞, as an ordinary number should be clear from the examples given above. For instance, if we put $n = \infty$ in the expression $(2n + 1)/(3n + 4)$, we would get $(2\infty + 1)/(3\infty + 4)$. Now, a multiple of ∞ is still ∞, and a number added to ∞ is still ∞, so we should get ∞/∞. Had ∞ been an ordinary number, subject to the ordinary rules of arithmetic, this expression would simply be equal to 1. But it is *not* equal to 1; it is 2/3, as we have seen. A similar situation arises when we try to "compute" $\infty - \infty$. It would be tempting to say that since any number when subtracted from itself gives 0, we should have $\infty - \infty = 0$. That this may be false can be seen from the expression $1/x^2 - [(\cos x)/x]^2$, where "cos" is the cosine function studied in trigonometry. As $x \to 0$, each of the two terms tends to infinity; yet, with the help of a little trigonometry, it can be shown that the entire expression approaches the limit 1.

Expressions such as ∞/∞ or $\infty - \infty$ are known as "indeterminate forms." These expressions have no preassigned value; they can be evaluated only through a limiting process. Loosely speaking, in every indeterminate form there is a "struggle" between two quantities, one tending to make the expression numerically large, the other tending to make it numerically small. The final outcome depends on the precise limiting process involved. The indeterminate forms most commonly encountered in mathematics are $0/0$, ∞/∞, $0 \cdot \infty$, $\infty - \infty$, 0^0, ∞^0, and 1^∞. It is to the last form that $(1 + 1/n)^n$ belongs.

In an indeterminate expression, algebraic manipulation alone may not be enough to determine the final outcome of the limiting process.

Of course, we could use a computer or a calculator to compute the expression for very large values of n, say a million or a billion. But such a computation can only *suggest* the limiting value. We have no assurance that this value will indeed hold up for a still larger n. This state of affairs underscores a fundamental difference between mathematics and the sciences that are based on experimental or observational evidence, such as physics and astronomy. In those sciences, if a certain result—say a numerical relation between the temperature of a given amount of gas and its pressure—is supported by a large number of experiments, that result may then be regarded as a law of nature.

A classic example is afforded by the universal law of gravitation, discovered by Isaac Newton and enunciated in his great work, *Philosophiae naturalis principia mathematica* (1687). The law says that any two material bodies—be they the sun and a planet revolving around it or two paperclips placed on the table—exert on each other a gravitational force proportional to the product of their masses and inversely proportional to the square of the distance between them (more precisely, between their centers of mass). For more than two centuries this law was one of the rock foundations of classical physics; every astronomical observation seemed to corroborate it, and it is still the basis for calculating the orbits of planets and satellites. It was only in 1916 that Newton's law of gravitation was replaced by a more refined law, Einstein's general theory of relativity. (Einstein's law differs from Newton's only for extremely large masses and speeds close to the speed of light.) Yet there is no way that Newton's law— or any other law of physics—can be proved in the mathematical sense of the word. A *mathematical* proof is a chain of logical deductions, all stemming from a small number of initial assumptions ("axioms") and subject to the strict rules of mathematical logic. Only such a chain of deductions can establish the validity of a mathematical law, a *theorem*. And unless this process has been satisfactorily carried out, no relation—regardless of how often it may have been confirmed by observation—is allowed to become a law. It may be given the status of a *hypothesis* or a *conjecture*, and all kinds of tentative results may be drawn from it, but no mathematician would ever base definitive conclusions on it.

As we saw in the last chapter, the expression $(1 + 1/n)^n$, for very large values of n, seems to approach the number 2.71828 as a limit. But in order to determine this limit with any certainty—or even to prove that the limit exists in the first place—we must use methods other than merely computing individual values. (Besides, it becomes increasingly difficult to compute the expression for large n's—one must use logarithms to do the exponentiation.) Fortunately, such a method is available, and it makes use of the *binomial formula*.

A *binomial* is any expression consisting of the sum of two terms; we may write such an expression as $a + b$. One of the first things we learn in elementary algebra is how to find successive powers of a binomial—how to expand the expression $(a + b)^n$ for $n = 0, 1, 2, \ldots$. Let us list the results for the first few n's:

$$
\begin{aligned}
(a + b)^0 &= 1 \\
(a + b)^1 &= a + b \\
(a + b)^2 &= a^2 + 2ab + b^2 \\
(a + b)^3 &= a^3 + 3a^2b + 3ab^2 + b^3 \\
(a + b)^4 &= a^4 + 4a^3b + 6a^2b^2 + 4ab^3 + b^4
\end{aligned}
$$

From these few examples it is easy to see the general pattern: the expansion of $(a + b)^n$ consists of $n + 1$ terms, each of the form $a^{n-k}b^k$, where $k = 0, 1, 2, \ldots, n$. Hence, as we go from left to right the exponent of a decreases from n to 0 (we can write the last term as a^0b^n), while that of b increases from 0 to n. The coefficients of the various terms, known as the *binomial coefficients*, form a triangular scheme:

This scheme is known as *Pascal's triangle*, after the French philosopher and mathematician Blaise Pascal (1623–1662), who used it in his theory of probability (the scheme itself had been known much earlier; see figs. 2, 3, and 4). In this triangle, each number is the sum of the two numbers immediately to its left and right *in the row above the number*. For example, the numbers in the fifth row, 1, 4, 6, 4, 1, are obtained from those in the fourth row as follows:

(Note that the coefficients are the same whether we start from left or right.)

There is one drawback in using Pascal's triangle to find the binomial coefficients: we must first compute all the rows above the one we are interested in, a process that becomes increasingly time-consuming as n increases. Fortunately, there is a formula that allows us to find these coefficients without depending on Pascal's triangle. If we denote the coefficient of the term $a^{n-k}b^k$ by ^nC_k, then

FIG. 2. Pascal's triangle appears on the title page of an arithmetic work by Petrus Apianus (Ingolstadt, 1527).

$$^nC_k = \frac{n!}{k!(n-k)!}.$$

(1)

The symbol $n!$, called *n factorial*, denotes the product $1 \cdot 2 \cdot 3 \cdot \ldots \cdot n$; the first few values of $n!$ are $1! = 1$, $2! = 1 \cdot 2 = 2$, $3! = 1 \cdot 2 \cdot 3 = 6$, and $4! = 1 \cdot 2 \cdot 3 \cdot 4 = 24$ (we also define $0!$ to be 1). If, for example, we apply this formula to the expansion of $(a + b)^4$, we get the coefficients $^4C_0 = 4!/(0! \cdot 4!) = 1$, $^4C_1 = 4!/(1! \cdot 3!) = 1 \cdot 2 \cdot 3 \cdot 4/(1 \cdot 2 \cdot 3) = 4$, $^4C_2 = 4!/(2! \cdot 2!) = 6$, $^4C_3 = 4!/(3! \cdot 1!) = 4$, and $^4C_4 = 4!/(4! \cdot 0!) = 1$—the same numbers that appear in the fifth row of Pascal's triangle.

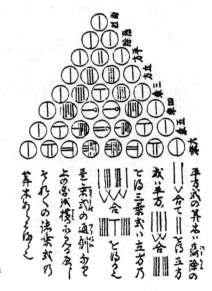

FIG. 3. Pascal's triangle in a Japanese work of 1781.

חכמת האלגעברא · הנשגבה

פרק יח מדרגות הנשגבות בכלל , סגולת סדר אבריהם וידותהם המהולל
בשם (בינאמישע [פורמל]) משפטי חלופי המצב מן הגיפים
(פערנוטטמליחן) , וחלופי הקשורים בהם (קאמבינאטיאנגן) ,

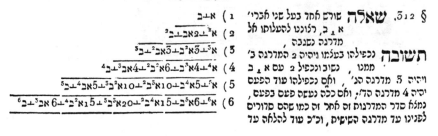

1) א—ב	§ 312. **שאלה** שורש אחד בעל שני אברי'
2) א²—2אב—ב²	א , ב , רצונט להעלותו אל
	מדרגה נשגבה ,
3) א³—3א²ב—3אב²—ב³	**תשובה** נכפילהו בעצמו ויהיה 2 מדרגה ב'
4) א⁴—4א³ב—6א²ב²—4אב³—ב⁴	ממנו , נשוב ונכפיל 2 עם א , ב
5) א⁵—5א⁴ב—10א³ב²—10א²ב³—5אב⁴—ב⁵	ויהיה 5 מדרגה הג' , ואם נכפילהו עוד הפעם
6) א⁶—6א⁵ב—15א⁴ב²—20א³ב³—15א²ב⁴—6אב⁵—ב⁶	יהיה 4 מדרגה הד'; ואם כלה נעשה פעם בפעם
	נמלא סדר המדרגות זה אחר זה כמו שהם סדורים
	לפנינו עד מדרגה השישים , וכ"ג עוד להלאה עד

FIG. 4. The expansion of $(a + b)^n$ for $n = 1, 2, 3, \ldots, 6$. From a Hebrew algebra book by Hayim Selig Slonimski (Vilnius, 1834). The formulas use Hebrew letters and are read from right to left.

The binomial formula can easily be proved for all positive integral values of n by the process known as mathematical induction: we show that if the formula is true for all values of n up to, say, m, then it must also be true for $n = m + 1$ (it is, of course, true for $n = 1$, since $(a + b)^1 = a + b$). We note that the expansion of $(a + b)^n$ comes to an end after exactly $n + 1$ terms. As we shall see in Chapter 8, one of the first great achievements of Isaac Newton was to extend this formula to the case where n is a negative integer or even a fraction; in these cases the expansion will involve an infinite number of terms.

A quick look at equation 1 will show that we can write it in an alternative form,

$$^nC_k = \frac{n \cdot (n - 1) \cdot (n - 2) \cdot \ldots \cdot (n - k + 1)}{k!}. \tag{2}$$

This is because $n! = 1 \cdot 2 \cdot 3 \cdot \ldots \cdot n$ while $(n - k)! = 1 \cdot 2 \cdot 3 \cdot \ldots \cdot (n - k)$, so that all the factors from 1 to $(n - k)$ in the numerator of equation 1 cancel with those in the denominator, leaving only the factors $n \cdot (n - 1) \cdot (n - 2) \cdot \ldots \cdot (n - k + 1)$. With equation 2 in mind, we can now apply the binomial formula to the expression $(1 + 1/n)^n$. We have $a = 1$ and $b = 1/n$, so that

$$\left(1 + \frac{1}{n}\right)^n = 1 + n \cdot \left(\frac{1}{n}\right) + \frac{n \cdot (n - 1)}{2!} \cdot \left(\frac{1}{n}\right)^2$$
$$+ \frac{n \cdot (n - 1) \cdot (n - 2)}{3!} \cdot \left(\frac{1}{n}\right)^3 + \ldots + \left(\frac{1}{n}\right)^n.$$

After a slight manipulation this becomes

$$\left(1 + \frac{1}{n}\right)^n = 1 + 1 + \frac{(1 - \frac{1}{n})}{2!} + \frac{(1 - \frac{1}{n}) \cdot (1 - \frac{2}{n})}{3!} + \ldots + \frac{1}{n^n}. \tag{3}$$

Since we are looking for the *limit* of $(1 + 1/n)^n$ as $n \to \infty$, we must let n increase without bound. Our expansion will then have more and more terms. At the same time, the expression within each pair of parentheses will tend to 1, because the limits of $1/n, 2/n, \ldots$ as $n \to \infty$ are all 0. We thus get

$$\lim_{n \to \infty} \left(1 + \frac{1}{n}\right)^n = 1 + 1 + \frac{1}{2!} + \frac{1}{3!} + \ldots. \tag{4}$$

We should add that even this derivation is not entirely sufficient to prove that the desired limit does indeed exist (a complete proof is found in Appendix 2). But for now let us accept the existence of this limit as a fact. Denoting the limit by the letter e (more about the choice of this letter later), we thus have

$$e = 2 + \frac{1}{2!} + \frac{1}{3!} + \frac{1}{4!} + \ldots. \tag{5}$$

Not only is it much easier to compute the terms of this *infinite series* and add as many of them as we please, but the sum will approach its limiting value much faster than when computing $(1 + 1/n)^n$ directly. The first seven partial sums of our series are:

$2 =$	2
$2 + 1/2 =$	2.5
$2 + 1/2 + 1/6 =$	$2.666\ldots$
$2 + 1/2 + 1/6 + 1/24 =$	$2.708333\ldots$
$2 + 1/2 + 1/6 + 1/24 + 1/120 =$	$2.716666\ldots$
$2 + 1/2 + 1/6 + 1/24 + 1/120 + 1/720 =$	$2.7180555\ldots$
$2 + 1/2 + 1/6 + 1/24 + 1/120 + 1/720 + 1/5{,}040 =$	$2.718253968\ldots$

We see that the terms of each sum decrease rapidly (this is because of the rapid growth of $k!$ in the denominator of each term), so that the series converges very fast. Moreover, since all terms are positive, the convergence is *monotone*: each additional term brings us closer to the limit (this is not so with a series whose terms have alternating signs). These facts play a role in the existence proof of $\lim_{n\to\infty} (1 + 1/n)^n$. For now, however, let us accept that e has the approximate value 2.71828 and that this approximation can be improved by adding more and more terms of the series, until the desired accuracy is reached.

NOTE

1. We exclude the trivial case where all the terms of the sequence are equal or where we artificially insert the limiting value as one of the members of the sequence. The definition of limit will, of course, hold for these cases as well.

Some Curious Numbers Related to e

$e^{-e} = 0.065988036\ldots$

Leonhard Euler proved that the expression $x^{x^{x^{x^{\cdot^{\cdot^{\cdot}}}}}}$, as the number of exponents grows to infinity, tends to a limit if x is between e^{-e} ($= 1/e^e$) and $e^{1/e}$.[1]

$e^{-\pi/2} = 0.207879576\ldots$

As Euler showed in 1746, the expression i^i (where $i = \sqrt{-1}$) has infinitely many values, all of them real: $i^i = e^{-(\pi/2 + 2k\pi)}$, where $k = 0, \pm1, \pm2, \ldots$. The principal value of these (the value for $k = 0$) is $e^{-\pi/2}$.

$1/e = 0.367879441\ldots$

The limit of $(1 - 1/n)^n$ as $n \to \infty$. This number is used to measure the rate of decay of the exponential function $y = e^{-at}$. When $t = 1/a$ we have $y = e^{-1} = 1/e$. It also appears in the "misplaced envelope" problem posed by Nicolaus Bernoulli: If n letters are to go in n addressed envelopes, what is the probability that every letter will be placed in a wrong envelope? As $n \to \infty$, the probability approaches $1/e$.[2]

$e^{1/e} = 1.444667861\ldots$

The solution of Jakob Steiner's problem: Find the maximum value attained by the function $y = x^{1/x} = \sqrt[x]{x}$. This value is attained when $x = e$.[3]

$878/323 = 2.718266254\ldots$

The closest rational approximation to e using integers below 1,000.[4] It is easy to memorize and is reminiscent of the rational approximation $355/113 = 3.14159292\ldots$ for π.

$e = 2.718281828\ldots$

The base of natural logarithms (also known as Napierian logarithms, although without historical justification) and the limit of $(1 + 1/n)^n$ as $n \to \infty$. The recurring block of digits 1828 is misleading, for e is an irrational number and is represented by a nonterminating, nonrepeating decimal. The irrationality of e was proved in 1737 by Euler. Charles Hermite in 1873 proved that e is transcendental; that is, it cannot be a solution of a polynomial equation with integer coefficients.

The number *e* can be interpreted geometrically in several ways. The area under the graph of $y = e^x$ from $x = -\infty$ to $x = 1$ is equal to *e*, as is the slope of the same graph at $x = 1$. The area under the hyperbola $y = 1/x$ from $x = 1$ to $x = e$ is equal to 1.

$e + \pi = 5.859874482 \ldots$
$e \cdot \pi = 8.539734223 \ldots$
These numbers rarely show up in applications; it is not known whether they are algebraic or transcendental.[5]

$e^e = 15.15426224 \ldots$
It is not known whether this number is algebraic or transcendental.[6]

$\pi^e = 22.45915772 \ldots$
It is not known whether this number is algebraic or transcendental.[7]

$e^\pi = 23.14069263 \ldots$
Alexandr Gelfond in 1934 proved that this number is transcendental.[8]

$e^{e^e} = 3,814,279.104 \ldots$
Note how much larger this number is than e^e. The next number in this progression, $e^{e^{e^e}}$, has 1,656,521 digits in its integral part.

✧ ✧ ✧

Two other numbers related to *e* are:

$\gamma = 0.577215664 \ldots$
This number, denoted by the Greek letter gamma, is known as Euler's constant; it is the limit of $1 + 1/1 + 1/2 + 1/3 + 1/4 + \ldots + 1/n - \ln n$ as $n \to \infty$. In 1781 Euler calculated this number to sixteen places. The fact that the limit exists means that although the series $1 + 1/2 + 1/3 + 1/4 + \ldots + 1/n$ (known as the harmonic series) diverges as $n \to \infty$, the difference between it and $\ln n$ approaches a constant value. It is not known whether γ is algebraic or transcendental, or even if it is rational or irrational.[9]

$\ln 2 = 0.693147181 \ldots$
This is the sum of the harmonic series with alternating signs, $1 - 1/2 + 1/3 - 1/4 + - \ldots$, obtained from Nicolaus Mercator's series $\ln (1 + x) = x - x^2/2 + x^3/3 - x^4/4 + - \ldots$ by substituting $x = 1$. It is the number to which *e* must be raised to get 2: $e^{0.693147181 \ldots} = 2$.

NOTES AND SOURCES

1. David Wells, *The Penguin Dictionary of Curious and Interesting Numbers* (Harmondsworth: Penguin Books, 1986), p. 35.

2. Ibid., p. 27. See also Heinrich Dörrie, *100 Great Problems of Elementary Mathematics: Their History and Solution*, trans. David Antin (New York: Dover, 1965), pp. 19–21.

3. Dörrie, *100 Great Problems*, p. 359.

4. Wells, *Dictionary of Curious and Interesting Numbers*, p. 46.

5. George F. Simmons, *Calculus with Analytic Geometry* (New York: McGraw-Hill, 1985), p. 737.

6. Carl B. Boyer, *A History of Mathematics*, rev. ed. (New York: John Wiley, 1989), p. 687.

7. Ibid.

8. Ibid.

9. Wells, *Dictionary of Curious and Interesting Numbers*, p. 28.

5

Forefathers of the Calculus

If I have seen further [than you and Descartes],

it is by standing upon the shoulders of Giants.

—SIR ISAAC NEWTON to Robert Hooke

Great inventions generally fall into one of two categories: some are the product of a single person's creative mind, descending on the world suddenly like a bolt out of the blue; others—by far the larger group—are the end product of a long evolution of ideas that have fermented in many minds over decades, if not centuries. The invention of logarithms belongs to the first group, that of the calculus to the second.

It is usually said that the calculus was invented by Isaac Newton (1642–1727) and Gottfried Wilhelm Leibniz (1646–1716) during the decade 1665–1675, but this is not entirely accurate. The central idea behind the calculus—to use the limit process to derive results about ordinary, finite objects—goes back to the Greeks. Archimedes of Syracuse (ca. 287–212 B.C.), the legendary scientist whose military inventiveness is said to have defied the Roman invaders of his city for more than three years, was one of the first to use the limit concept to find the area and volume of various planar shapes and solids. For reasons that we shall soon see, he never used the term *limit*, but that is precisely what he had in mind.

Elementary geometry allows us to find the perimeter and area of any triangle, and hence of any polygon (a closed planar shape made up of straight line segments). But when it comes to curved shapes, elementary geometry is powerless. Take the circle as an example. In beginning geometry we learn that the circumference and area of a circle are given by the simple formulas $C = 2\pi r$ and $A = \pi r^2$, respectively. But the seeming simplicity of these formulas is misleading, for the constant π appearing in them—the ratio of the circumference of a circle to its diameter—is one of the most intriguing numbers in mathematics. Its nature was not fully established until late in the nineteenth century, and even today some questions about it remain unanswered.

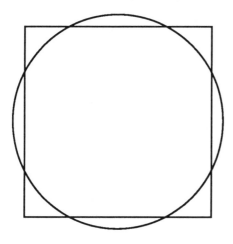

FIG. 5. According to the Rhind Papyrus (ca. 1650 B.C.), a circle has the same area as a square whose side is 8/9 the diameter of the circle.

The *value* of π has been known with remarkable accuracy for a long time. An Egyptian text dating to 1650 B.C. and known as the Rhind Papyrus (named after the Scottish Egyptologist A. Henry Rhind, who purchased it in 1858) contains the statement that a circle has the same area as a square whose side is 8/9 the diameter of the circle (fig. 5). If we denote the diameter by d, the statement translates into the equation $\pi(d/2)^2 = [(8/9)d]^2$, from which we get, after canceling d^2, $\pi/4 = 64/81$, or $\pi = 256/81 \approx 3.16049$.[1] This result comes within 0.6 percent of the true value of π (3.14159, rounded to five decimal places)—remarkably accurate for a text written almost four thousand years ago![2]

Over the centuries many values for π have been given. But up to the Greek era, all these values were essentially empirical: they were obtained by actually measuring the circumference of a circle and dividing it by the diameter. It was Archimedes who first proposed a method that could give the value of π to any desired accuracy by a *mathematical procedure*—an algorithm—rather than by measurement.

Archimedes' idea was to take a circle and inscribe in it a series of regular polygons of more and more sides. (In a regular polygon, all sides are equal in length and all angles have the same measure.) Each polygon has a perimeter slightly less than the circumference of the circle; but as we increase the number of sides, the polygons will approach the circle closer and closer (fig. 6). By finding the perimeter of each polygon and dividing it by the diameter, we obtain an approximation for π, and this approximation can be improved by simply increasing the number of sides. Now, because the inscribed polygons approach the circle from within, all these approximations will fall short of the true value π. Archimedes therefore repeated the process with *circumscribing* polygons (fig. 7), giving him a series of approx-

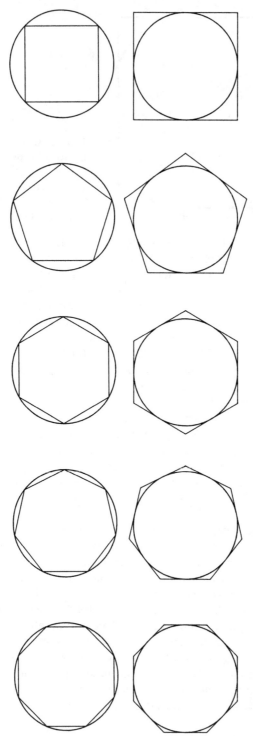

FIG. 6 (*left*). Regular polygons inscribed in a circle.

FIG. 7 (*right*). Regular polygons circumscribing a circle.

imations that exceed π. For any given number of sides, then, the true value of π is "squeezed" between a lower and an upper bound; by increasing the number of sides, we can make the interval between these bounds as narrow as we please, like the jaws of a vise closing on each other. Using inscribed and circumscribing polygons of ninety-six sides (which he arrived at by starting with a regular hexagon and repeatedly doubling the number of sides), Archimedes calculated the value of π to be between 3.14103 and 3.14271—an accuracy that even today is sufficient for most practical purposes.[3] If we could circumscribe the equator of a twelve-inch diameter globe with a polygon of ninety-six sides, the corners would be barely noticeable over the globe's smooth surface.

Archimedes' achievement was a milestone in the history of mathematics, but he did not stop there. He was equally interested in another common figure, the parabola—approximately the curve traced by a stone thrown into the air (the trajectory would be an exact parabola if there were no air to resist the motion). The parabola shows up in a host of applications. The large dish antennas used in modern telecommunication have a parabolic cross section, as do the silvered reflecting surfaces of a car's headlights. Archimedes' interest in the parabola may have stemmed from a certain property of this curve: its ability to reflect rays of light coming from infinity and concentrate them at a single point, the *focus* (the word in Latin means "fireplace"). He is said to have built huge parabolic mirrors, which he aimed at the Roman fleet besieging his city, so that the sun's rays, converging at the focus of each parabola, would set the enemy ships ablaze.

Archimedes also investigated the more theoretical aspects of the parabola, in particular, how to find the area of a parabolic sector. He solved this problem by dividing the sector into a series of triangles whose areas decrease in a geometric progression (fig. 8). By continuing this progression on and on, he could make the triangles fit the parabola as closely as he pleased—"exhaust" it, so to speak. When he added the areas of the individual triangles (using the formula for the sum of a geometric progression), Archimedes found that the total area approached 4/3 the area of triangle ABC; more precisely, by taking more and more triangles, he could make the total area as close to this value as he pleased.[4] In modern terms, the area of the triangles approaches the *limit* 4/3 (taking the area of triangle ABC as 1) as the number of triangles tends to infinity. Archimedes, however, was careful to formulate his solution in terms of finite sums only; the word *infinity* never appeared in his argument, and for a good reason: the Greeks banned infinity from their discussions and refused to incorporate it into their mathematical system. We shall soon see why.

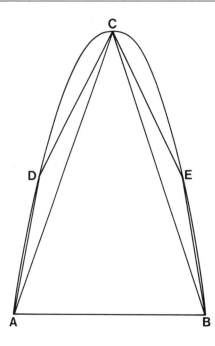

FIG. 8. Archimedes' method of exhaustion applied to a parabola.

Archimedes' method came to be known as the *method of exhaustion*. Although it did not originate with him (its invention is attributed to Eudoxus, around 370 B.C.), he was the first to apply it successfully to the parabola. He could not make it work, however, in the case of two other famous curves, the ellipse and the hyperbola, which, together with the parabola, make up the family of *conic sections*.[5] Despite repeated attempts, Archimedes could not find the area of the elliptic and hyperbolic sectors, although he did guess correctly that the area of the entire ellipse is πab (where a and b are the lengths of the major and minor axes). These cases had to wait for the invention of integral calculus two thousand years later.

Area of ellipse! →

The method of exhaustion came very close to our modern integral calculus. Why, then, did the Greeks fail to discover the calculus? There were two reasons: the Greeks' uneasiness with the concept of infinity—what has been called their *horror infiniti*—and the fact that they did not possess the language of algebra. Let us begin with the second reason. The Greeks were masters of geometry—virtually all of classical geometry was developed by them. Their contribution to algebra, however, was marginal. Algebra is essentially a language, a collection of symbols and a set of rules with which to operate with these symbols. To develop such a language, one must have a good system of notation, and here the Greeks failed. Their failure can be attributed to their static view of the world, and of geometry in particular: they considered all geometric quantities as having fixed, given magnitudes. Our modern practice of labeling a quantity by a single letter, say x, and regarding it as a variable that can assume a range of values was alien to them. The Greeks denoted the line segment from A to B by AB, a rectangle with vertices A, B, C, D by $ABCD$, and so on. Such a system of notation served quite well its intended purpose of establishing the host of relations that exist among the various parts of a figure—the body of theorems that make up classical geometry. But when it came to expressing relations among *variable* quantities, the system was woefully inadequate. To express such relations efficiently, one must resort to the language of algebra.

The Greeks were not entirely ignorant of algebra. Many of the formulas of elementary algebra were known to them, but these were always understood to represent geometric relations among various parts of a figure. To begin with, a number was interpreted as the length of a line segment, the sum of two numbers as the combined length of two segments placed end to end along the same straight line, and the product of two numbers as the area of a rectangle with these segments as sides. The familiar formula $(x + y)^2 = x^2 + 2xy + y^2$ could then be interpreted in the following way: along a straight line, draw a segment of length $AB = x$; at its endpoint draw a second segment of length $BC = y$, and construct a square of side $AC = x + y$, as in figure 9. This square can be dissected into four parts: two small squares with areas $AB \cdot AB = x^2$ and $BC \cdot BC = y^2$, and two rectangles with area $AB \cdot BC = xy$. (There are some subtleties in this proof, such as the fact that the rectangles $BCDE$ and $EFGH$ are congruent and hence have the same area; the Greeks took great pain to account for all these details, meticulously justifying every step in the proof.) Similar methods were used to prove other algebraic relations, such as $(x - y)^2 = x^2 - 2xy + y^2$ and $(x + y)(x - y) = x^2 - y^2$.

One cannot but marvel at the Greeks' success in establishing a

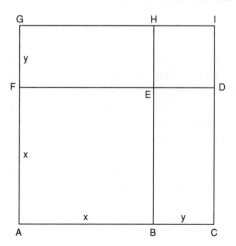

FIG. 9. Geometric proof of the formula $(x + y)^2 = x^2 + 2xy + y^2$.

large part of elementary algebra by geometric means alone. But this "geometric algebra" could not be used as an efficient, workable mathematical tool. Lacking a good system of notation—an algebra in the modern sense of the word—the Greeks were deprived of its single greatest advantage: its ability to express in a concise way relations among variable quantities. And that included the concept of infinity.

Because it is not a real number, infinity cannot be dealt with in a purely numerical sense. We have already seen that in order to find the value of various indeterminate forms one must use a limiting process, which in turn requires a good deal of algebraic skill. Without such skill, the Greeks could not properly deal with infinity. As a result, they avoided it, even feared it. In the fourth century B.C. the philosopher Zeno of Elea came up with four paradoxes—or "arguments" as he called them—whose purpose was to demonstrate the inability of mathematics to cope with the concept of infinity. One of his paradoxes purports to show that motion is impossible: in order for a runner to move from point A to point B, he must first reach the midpoint of AB, then the midpoint of the remaining distance, and so on *ad infinitum* (fig. 10). Since this process requires an infinite number of steps, Zeno argued, the runner will never reach his destination.

It is easy to explain the runner's paradox using the limit concept. If we take the line segment AB to be of unit length, then the total distance the runner must cover is given by the infinite geometric series $1/2 + 1/4 + 1/8 + 1/16 + \ldots$. This series has the property that no matter how many terms we add, its sum will never reach 1, let alone exceed 1; yet we can make the sum get as close to 1 as we please simply by adding more and more terms. We say that the series *converges* to 1, or has the *limit* 1, as the number of terms tends to infinity. Thus the runner will cover a total distance of exactly one unit (the length of the original distance AB), and the paradox is settled. The Greeks, however, found it difficult to accept the fact that an infinite sum of numbers may converge to a finite limit. The thought of going out to infinity was taboo to them. This is why Archimedes, in his method of exhaustion, never mentioned the word *infinity*. If he had an infinite process in mind—and there can be little doubt that he had—he was careful to formulate it as a finite process that could be repeated again and again until the desired accuracy was achieved.[6] Consequently, the method of exhaustion, while being a model of rigorous thinking, was so encumbered with pedantic details as to make it practically useless in dealing with all but the most simple geometric shapes. What is more, the answer to any specific problem had to be known in advance; only then could the method of exhaustion be used to establish the result rigorously.[7]

Thus, while Archimedes had a firm intuitive grasp of the limit concept, he could not make the crucial step of transforming it into a

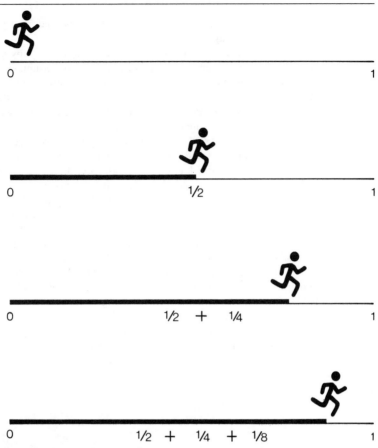

FIG. 10. The runner's paradox.

general and systematic procedure—an algorithm—that could be applied to a variety of different cases. Like Moses gazing on the Promised Land from Mount Nebo but not allowed to enter it, he came close to discovering a new science,[8] but then had to pass the torch to his successors.

NOTES AND SOURCES

1. The value 256/81 can be neatly written as $(4/3)^4$.

2. *The Rhind Mathematical Papyrus*, trans. Arnold Buffum Chace (Reston, Va.: National Council of Teachers of Mathematics, 1978), problems 41–43 and 50. The Rhind Papyrus is now in the British Museum.

3. Ronald Calinger, ed., *Classics of Mathematics* (Oak Park, Ill.: Moore Publishing Company, 1982), pp. 128–131.

4. Ibid., pp. 131–133.

5. The conic sections also include the circle and a pair of straight lines;

these, however, are merely special cases of the ellipse and the hyperbola. We will have more to say about the conic sections later.

6. Thus, in the case of the parabola, Archimedes proved by a double *reductio ad absurdum* (an indirect proof that starts by assuming that the assertion to be proved is wrong and then produces a contradiction) that the sum of the infinite series $1 + 1/4 + 1/4^2 + \ldots$ can be neither greater than nor less than 4/3, and must therefore be equal to 4/3. Today, of course, we would use the formula for the sum of an infinite geometric series, $1 + q + q^2 + \ldots = 1/(1 - q)$, where $-1 < q < 1$, to obtain the result $1/(1 - 1/4) = 4/3$.

7. That Archimedes had a way of "guessing" such results in advance is confirmed in his treatise known as *The Method*, discovered in 1906 when J. L. Heiberg found a medieval manuscript in Constantinople whose text had been written over a much older and partially washed-out text. The older text turned out to be a tenth-century copy of several of Archimedes' works, among them *The Method*, long thought to have been forever lost. Thus the world was allowed a rare glimpse into Archimedes' thought process—an invaluable opportunity, since the Greeks, in proving their geometric theorems, did not leave any indication as to how these had been discovered. See Thomas L. Heath, *The Works of Archimedes* (1897; rpt. New York: Dover, 1953); this edition contains a 1912 supplement, "*The Method* of Archimedes," with an introductory note.

8. On this subject see Heath, *The Works of Archimedes*, ch. 7 ("Anticipations by Archimedes of the Integral Calculus").

6

Prelude to Breakthrough

Infinities and indivisibles transcend our finite
understanding, the former on account of their magnitude,
the latter because of their smallness; Imagine what they are
when combined.
—GALILEO GALILEI as Salviati in *Dialogues Concerning*
Two New Sciences (1638)[1]

About eighteen hundred years after Archimedes, a French mathematician by the name of François Viète (or Vieta, 1540–1603), in the course of his work in trigonometry, found a remarkable formula involving the number π:

$$\frac{2}{\pi} = \frac{\sqrt{2}}{2} \cdot \frac{\sqrt{2 + \sqrt{2}}}{2} \cdot \frac{\sqrt{2 + \sqrt{2 + \sqrt{2}}}}{2} \cdots$$

The discovery of this *infinite product* in 1593 marked a milestone in the history of mathematics: it was the first time an infinite process was explicitly written as a mathematical formula. Indeed, the most remarkable feature of Viète's formula, apart from its elegant form, is the three dots at the end, telling us to go on and on . . . *ad infinitum.* It shows that π can be found, at least in principle, by repeatedly using four operations of elementary mathematics—addition, multiplication, division, and square root extraction—all applied to the number 2.

Viète's formula broke an important psychological barrier, for the mere act of writing the three dots at its end signaled the acceptance of infinite processes into mathematics and opened the way to their widespread use. Soon to follow were other formulas involving infinite processes. The English mathematician John Wallis (1616–1703), whose work *Arithmetica infinitorum* (1655) would later influence young Newton, discovered another infinite product involving π:

$$\frac{\pi}{2} = \frac{2}{1} \cdot \frac{2}{3} \cdot \frac{4}{3} \cdot \frac{4}{5} \cdot \frac{6}{5} \cdot \frac{6}{7} \cdots$$

And in 1671, the Scotsman James Gregory (1638–1675) discovered the *infinite series*

$$\frac{\pi}{4} = \frac{1}{1} - \frac{1}{3} + \frac{1}{5} - \frac{1}{7} + - \ldots .$$

What makes these formulas so remarkable is that the number π, originally defined in connection with the circle, can be expressed in terms of integers alone, albeit through an infinite process. To this day, these formulas are among the most beautiful in all of mathematics.

But for all their beauty, the usefulness of these formulas as a means to compute π is rather limited. As we have seen, several good approximations to π were already known in ancient times. Over the centuries numerous attempts were made to arrive at ever better approximations, that is, to find the value of π correct to more and more decimal places. The hope was that the decimal expansion of π would eventually come to an end (that is, contain only zeros from a certain point on) or begin to repeat in a cycle. Either eventuality would mean that π is a *rational number*, a ratio of two integers (we know today that no such ratio exists and that π has a nonterminating, nonrepeating expansion). Among the many mathematicians who hoped to achieve this goal, one name is particularly noteworthy. Ludolph van Ceulen (1540–1610), a German-Dutch mathematician, devoted most of his productive life to the task of computing π, and in the last year of his life he arrived at a value correct to thirty-five places. So highly was this feat regarded at the time that his number is said to have been engraved on his tombstone in Leiden, and for many years German textbooks referred to π as the "Ludolphine number."[2] His accomplishment, however, did not shed any new light on the nature of π (van Ceulen simply repeated Archimedes' method with polygons of more sides), nor did it contribute anything new to mathematics in general.[3] Fortunately for mathematics, such a folly would not be repeated with e.

Thus, the newly discovered formulas were remarkable not so much for their practicality as for the insight they afforded into the nature of the infinite process. Here we have a good example of the different philosophies of two schools of mathematical thinking: the "pure" school versus the "applied." Pure mathematicians pursue their profession with little concern for practical applications (some even claim that the more removed mathematics is from practical matters, the better for the profession). To some members of this school, mathematical research is much like a good game of chess, an activity whose main reward is the intellectual stimulus it affords; others pursue their research for the freedom it allows, freedom to create one's own definitions and rules and erect on them a structure held together solely by the rules of mathematical logic. Applied mathematicians, by contrast, are more concerned with the vast range of problems arising from science and technology. They do not enjoy the same degree of freedom as their "pure" counterparts, for they are bound by the laws of nature governing the phenomenon under investigation. Of course, the dividing line between the two schools is not always so clear-cut: a "pure" field of research has often found some unexpected practical application (one example is the application of number theory to encoding and decoding classified messages), and conversely, applied problems have led to theoretical discoveries of the highest rank. Moreover, some of the greatest names in the history of mathematics—among them Archimedes, Newton, and Gauss—were equally eminent in both domains. Still, the dividing line is quite real and has become even more pronounced in our time, when narrow specialization has replaced the universalism of previous generations.

Over the years the dividing line between the two schools has shifted back and forth. In ancient, pre-Greek times, mathematics was entirely a practical vocation, created to deal with such mundane matters as mensuration (the measurement of area, volume, and weight), monetary questions, and the reckoning of time. It was the Greeks who transformed mathematics from a practical profession into an intellectual one, where knowledge for its own sake became the main goal. Pythagoras, who founded his famous school of philosophy in the sixth century B.C., embodied the ideals of pure mathematics at their highest. His inspiration came from the order and harmony of nature—not just the immediate nature around us, but the entire universe. The Pythagoreans believed that numbers are the prime cause behind everything in the world, from the laws of musical harmony to the motion of the planets. "Number rules the universe" was their motto, and by "number" they meant natural numbers and their ratios; everything else—negative numbers, irrational numbers, and even zero—was excluded. In the Pythagorean philosophy, numbers assumed an almost sacred status, with all kinds of mythical meanings attached to them; whether these numbers actually described the real world was irrelevant. As a result, Pythagorean mathematics was an esoteric, aloof subject removed from daily matters and put on an equal footing with philosphy, art, and music. Indeed, Pythagoras devoted much of his time to the laws of musical harmony. He is said to have devised a musical scale based on the "perfect" proportions of 2 : 1 (the octave), 3 : 2 (the fifth) and 4 : 3 (the fourth). Never mind that the laws of acoustics demanded a more complicated arrangement of notes; the important thing was that the scale rested on simple mathematical ratios.[4]

The Pythagorean philosophy exercised an enormous influence on generations of scientists for more than two thousand years. But as Western civilization began to emerge from the Middle Ages, emphasis shifted once again to applied mathematics. Two factors contributed to this shift: the great geographical discoveries of the fifteenth and sixteenth centuries brought within reach faraway lands waiting to be explored (and later exploited), and this in turn called for the development of new and improved navigational methods; and Copernicus's heliocentric theory forced scientists to reexamine earth's place in the universe and the physical laws that govern its motion. Both developments required an ever increasing amount of practical mathematics, particularly in spherical trigonometry. Thus the next two centuries brought to prominence a line of applied mathematicians of the first rank, starting with Copernicus himself and culminating with Kepler, Galileo, and Newton.

To Johannes Kepler (1571–1630), one of the strangest men in the history of science, we owe the discovery of the three planetary laws that bear his name. These he found after years of futile searches that led him first to the laws of musical harmony, which he believed govern the motion of the planets (whence came the phrase "music of the spheres"), and then to the geometry of the five Platonic solids,[5] which, according to him, determined the gaps between the orbits of the six known planets. Kepler was the perfect symbol of the period of transition from the old world to the new: he was at once an applied mathematician of the highest rank and an ardent Pythagorean, a mystic who was led (or misled) by metaphysical considerations as much as by sound scientific reasoning (he actively practiced astrology even as he made his great astronomical discoveries). Today Kepler's nonscientific activities, like those of his contemporary Napier, have largely been forgotten, and his name is secured in history as the founder of modern mathematical astronomy.

The first of Kepler's laws says that the planets move around the sun in ellipses, with the sun at the focus of each ellipse. This discovery sounded the final death knell to the old Greek picture of a geocentric universe in which the planets and stars were embedded in crystalline spheres that revolved around the earth once every twenty-four hours. Newton would later show that the ellipse (with the circle as a special case) is only one member of a family of orbits in which a celestial body can move, the others being the parabola and the hyperbola. These curves (to which we should add a pair of straight lines as a limiting case of a hyperbola) constitute the family of *conic sections*, so called because they can all be obtained by cutting a circular cone with a plane at various angles of incidence (fig. 11). The conic sections were already known to the Greeks, and Archimedes' contemporary Apollonius (ca. 260–190 B.C.) wrote an extensive treatise on them. Now, two thousand years later, the attention of mathematicians was once again focused on the conic sections.

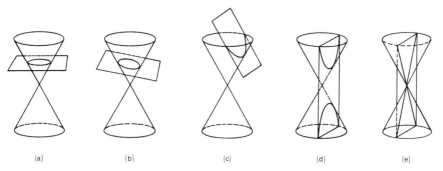

(a)   (b)   (c)   (d)   (e)

FIG. 11. The five conic sections.

Kepler's second law, the law of areas, states that the line joining a planet with the sun sweeps equal areas in equal times. Thus the question of finding the area of an elliptic segment—and more generally, of any conic section—suddenly became crucial. As we have seen, Archimedes had successfully used the method of exhaustion to find the area of a parabolic segment, but he failed with the ellipse and the hyperbola. Kepler and his contemporaries now showed a renewed interest in Archimedes' method; but whereas Archimedes was careful to use only finite processes—he never explicitly used the notion of infinity—his modern followers did not let such pedantic subtleties stand in their way. They took the idea of infinity in a casual, almost brazen manner, using it to their advantage whenever possible. The result was a crude contraption that had none of the rigor of the Greek method but that somehow seemed to work: the *method of indivisibles.* By thinking of a planar shape as being made up of an infinite number of infinitely narrow strips, the so-called "indivisibles," one can find the area of the shape or draw some other conclusions about it. For example, one can prove (*demonstrate* would be a better word) the relation between the area of a circle and its circumference by thinking of the circle as the sum of infinitely many narrow triangles, each with its vertex at the center and its base along the circumference (fig. 12). Since the area of each triangle is half the product of its base and height, the total area of all the triangles is half the product of their common height (the radius of the circle) and the sum of their bases (the circumference). The result is the formula $A = Cr/2$.

Of course, to derive this formula by the method of indivisibles is exercising wisdom at hindsight, since the formula had been known in antiquity (it can be obtained simply by eliminating π between the equations $A = \pi r^2$ and $C = 2\pi r$). Moreover, the method was flawed in several respects: To begin with, no one understood exactly what these "indivisibles" were, let alone how to operate with them. An indivisible was thought to be an infinitely small quantity—indeed, a

Hardly crude!

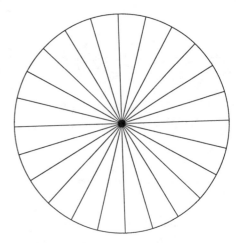

FIG. 12. The area of a circle can be thought of as the sum of infinitely many small triangles, each with its vertex at the center and base along the circumference.

quantity of magnitude 0—and surely if we add up any number of these quantities, the result should still be 0 (we recognize here the indeterminate expression $\infty \cdot 0$). Second, the method—if it worked at all—required a great deal of geometric ingenuity, and one had to devise the right kind of indivisibles for each problem. Yet, for all its flaws, the method somehow did work and in many cases actually produced new results. Kepler was one of the first to make full use of it. For a while he put aside his astronomical work to deal with a down-to-earth problem: to find the volume of various wine barrels (reportedly he was dissatisfied with the way wine merchants gauged the content of their casks). In his book *Nova stereometria doliorum vinariorum* (New solid geometry of wine barrels, 1615) Kepler applied the method of indivisibles to find the volumes of numerous solids of revolution (solids obtained by revolving a planar shape about an axis in the plane of the shape). He did this by extending the method to three dimensions, regarding a solid as a collection of infinitely many thin slices, or laminae, and summing up their individual volumes. In employing these ideas, he came within one step of our modern integral calculus.

NOTES AND SOURCES

1. Translated by Henry Crew and Alfonso De Salvio (1914; rpt. New York: Dover, 1914).

2. Petr Beckmann, *A History of π* (Boulder, Colo.: Golem Press, 1977), p. 102.

3. Van Ceulen's record has long since been broken. In 1989 two American researchers at Columbia University, using a supercomputer, calculated π to

480 million decimal places. Their number would stretch for some 600 miles if printed. See also Beckmann, *A History of* π, ch. 10.

4. Much of what we know about Pythagoras comes from works by his followers, often written centuries after his death; hence many of the "facts" about his life must be taken with a grain of salt. We will say more about Pythagoras in Chapter 15.

5. In a regular or Platonic solid all faces are regular polygons, and the same number of edges meet at each vertex. There are exactly five Platonic solids: the tetrahedron (four faces, each an equilateral triangle), the cube, the octahedron (eight equilateral triangles), the dodecahedron (twelve regular pentagons), and the icosahedron (twenty equilateral triangles). All five were known to the Greeks.

Indivisibles at Work

$\mathbf{A}$s an example of the method of indivisibles, let us find the area under the parabola $y = x^2$ from $x = 0$ to $x = a$. We think of the required region as made up of a large number n of vertical line segments ("indivisibles") whose heights y vary with x according to the equation $y = x^2$ (fig. 13). If these line segments are separated by a fixed horizontal distance d, their heights are d^2, $(2d)^2$, $(3d)^2$, ... , $(nd)^2$. The required area is thus approximated by the sum

$$[d^2 + (2d)^2 + (3d)^2 + \ldots + (nd)^2] \cdot d$$
$$= [1^2 + 2^2 + 3^2 + \ldots + n^2] \cdot d^3.$$

Using the well-known summation formula for the squares of the integers, this expression is equal to $[n(n + 1)(2n + 1)/6] \cdot d^3$, or after a slight algebraic manipulation, to

$$\frac{(1 + \frac{1}{n})(2 + \frac{1}{n})(nd)^3}{6}.$$

Since the length of the interval from $x = 0$ to $x = a$ is a, we have $nd = a$, so that the last expression becomes

$$\frac{(1 + \frac{1}{n})(2 + \frac{1}{n})a^3}{6}.$$

Finally, if we let the number of indivisibles grow without bound (that is, let $n \to \infty$), the terms $1/n$ and $2/n$ will tend to 0, and we get as our area

$$A = \frac{a^3}{3}.$$

This, of course, agrees with the result $A = \int_0^a x^2 dx = a^3/3$ obtained by integration. It is also compatible with Archimedes' result, obtained by the method of exhaustion, that the area of the parabolic segment OPQ in figure 13 is 4/3 the area of the triangle OPQ, as can easily be checked.

Aside from the fact that the pioneers of the method of indivisibles were not clear about what exactly an "indivisible" is, the method is crude and depends heavily on finding some suitable summation formula. For example, it cannot be used to find the area under the hyper-

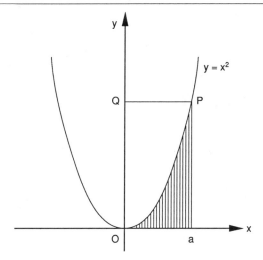

FIG. 13. Finding the area under a parabola by the method of indivisibles.

bola $y = 1/x$ because there is no summation formula for the reciprocals of the integers. Thus, while the method works for many particular cases, it lacks the generality and algorithmic nature of the modern technique of integration.

7

Squaring the Hyperbola

Grégoire Saint-Vincent is the greatest of circle-squarers . . .
he found the property of the area of the hyperbola which
led to Napier's logarithms being called hyperbolic.
—Augustus De Morgan, *The Encyclopedia of*
Eccentrics (1915)

The problem of finding the area of a closed planar shape is known as *quadrature*, or squaring. The word refers to the very nature of the problem: to express the area in terms of units of area, that is, squares. To the Greeks this meant that the given shape had to be transformed into an equivalent one whose area could be found from fundamental principles. To give a simple example, suppose we want to find the area of a rectangle of sides a and b. If this rectangle is to have the same area as a square of side x, we must have $x^2 = ab$, or $x = \sqrt{(ab)}$. Using a straightedge and compass, we can easily construct a segment of length $\sqrt{(ab)}$, as shown in figure 14. Thus we can affect the quadrature of any rectangle, and hence of any parallelogram and any triangle, because these shapes can be obtained from a rectangle by simple constructions (fig. 15). The quadrature of any polygon follows immediately, because a polygon can always be dissected into triangles.

In the course of time, this purely geometric aspect of the problem of quadrature gave way to a more computational approach. The actual construction of an equivalent shape was no longer considered necessary, so long as it could be demonstrated that such a construction could be done *in principle*. In this sense the method of exhaustion was not a true quadrature, since it required an infinite number of steps and thus could not be achieved by purely geometric means. But with the admission of infinite processes into mathematics around 1600, even this restriction was dropped, and the problem of quadrature became a purely computational one.

Among the shapes that have stubbornly resisted all attempts at squaring was the hyperbola. This curve is obtained when a cone is cut

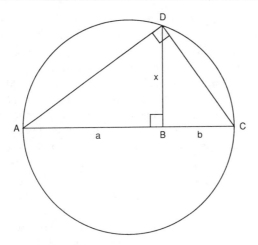

FIG. 14. Constructing a segment of length $x = \sqrt{ab}$ with straightedge and compass. On a line lay a segment AB of length a, at its end lay a second segment BC of length b, and construct a semicircle with AC as diameter. At B erect a perpendicular to AC and extend it until it meets the circle at D. Let the length of BD be x. By a well-known theorem from geometry, $\sphericalangle ADC$ is a right angle. Hence $\sphericalangle BAD = \sphericalangle BDC$, and consequently triangles BAD and BDC are similar. Thus $AB/BD = BD/BC$ or $a/x = x/b$, from which we get $x = \sqrt{ab}$.

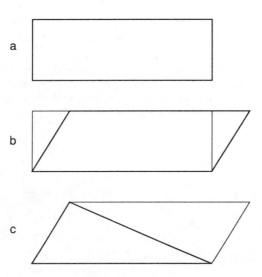

FIG. 15. The rectangle (a) and parallelogram (b) have the same area. The triangle (c) has half this area.

by a plane at an angle greater than the angle between the base of the cone and its side (hence the prefix "hyper," meaning "in excess of"). Unlike the familiar ice cream cone, however, here we think of a cone as having *two* nappes joined at the apex; as a result, a hyperbola has two separate and symmetric branches (see fig. 11[d]). In addition, the hyperbola has a pair of straight lines associated with it, namely, its two tangent lines at infinity. As we move along each branch outward from the center, we approach these lines ever closer but never reach them. These lines are the *asymptotes* of the hyperbola (the word in Greek means "not meeting"); they are the geometric manifestation of the limit concept discussed earlier.

The Greeks studied the conic sections from a purely geometric point of view. But the invention of analytic geometry in the seventeenth century made the study of geometric objects, and curves in particular, increasingly a part of algebra. Instead of the curve itself, one considered the *equation* relating the x and y coordinates of a point on the curve. It turns out that each of the conic sections is a special case of a *quadratic* (second-degree) equation, whose general form is $Ax^2 + By^2 + Cxy + Dx + Ey = F$. For example, if $A = B = F = 1$ and $C = D = E = 0$ we get the equation $x^2 + y^2 = 1$, whose graph is a circle with center at the origin and radius 1 (the unit circle). The hyperbola shown in figure 16 corresponds to the case $A = B = D = E = 0$ and $C = F = 1$; its equation is $xy = 1$ (or equivalently, $y = 1/x$), and its asymptotes are the x and y axes. Because its asymptotes are perpendicular to each other, this particular hyperbola is known as a *rectangular hyperbola*.

As we recall, Archimedes tried unsuccessfully to square the hyperbola. When the method of indivisibles was developed early in the seventeenth century, mathematicians renewed their attempts to achieve this goal. Now the hyperbola, unlike the circle and ellipse, is

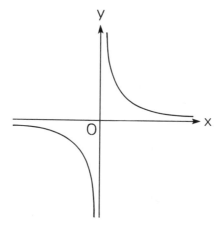

FIG. 16. The rectangular hyperbola $y = 1/x$.

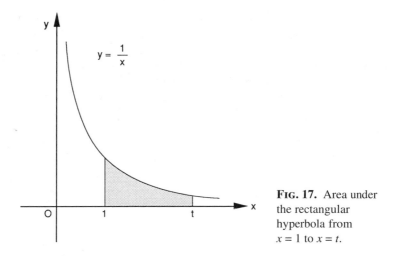

FIG. 17. Area under the rectangular hyperbola from $x = 1$ to $x = t$.

a curve that goes to infinity, so we must clarify what is meant by quadrature in this case. Figure 17 shows one branch of the hyperbola $xy = 1$. On the x-axis we mark the fixed point $x = 1$ and an arbitrary point $x = t$. By the *area under the hyperbola* we mean the area between the graph of $xy = 1$, the x-axis, and the vertical lines (ordinates) $x = 1$ and $x = t$. Of course, the numerical value of this area will still depend on our choice of t and is therefore a function of t. Let us denote this function by $A(t)$. The problem of squaring the hyperbola amounts to finding this function, that is, expressing the area as a formula involving the variable t.

Around the beginning of the seventeenth century several mathematicians independently attempted to solve this problem. Notable among them were Pierre de Fermat (1601–1665) and René Descartes (1596–1650), who, together with Blaise Pascal (1623–1662), form the great French triumvirate of mathematicians in the years just before the invention of the calculus. Like Bach and Handel in music, Descartes and Fermat are often mentioned together as sort of mathematical twins. However, apart from the fact that both were French and almost exact contemporaries, one could hardly find two figures less alike. Descartes began his professional life as a soldier, seeing action in many of the regional wars that raged throughout Europe in those days. He changed his allegiance more than once, switching to whichever side needed his services. Then one night he had a vision that God entrusted him with the key to unlocking the secrets of the universe. While still on military duty, he turned to philosophy and soon became one of the most influential philosophers of Europe. His motto, "I think, therefore I am," summarized his belief in a rational world governed by reason and mathematical design. His interest in mathematics, though, was secondary to his philosphical preoccupations. He published only one significant mathematical work—but that work changed the course of mathematics. In his *La Géométrie*, published in 1637 as one of three appendixes to his main philosophical work, *Discours de la méthode pour bien conduire sa raison et chercher la vérité dans les sciences* (Discourse on the method of reasoning well and seeking truth in the sciences), he introduced analytic geometry to the world.

The key idea of analytic geometry—said to have occurred to Descartes while he lay in bed late one morning and watched a fly move across the ceiling—was to describe every point in the plane by two numbers, its distances from two fixed lines (fig. 18). These numbers,

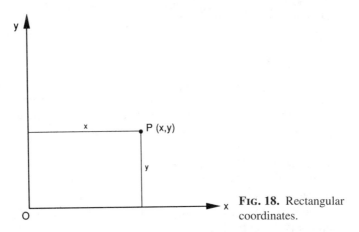

FIG. 18. Rectangular coordinates.

the *coordinates* of the point, enabled Descartes to translate geometric relations into algebraic equations. In particular, he regarded a curve as the locus of points having a given common property; by considering the coordinates of a point on the curve as variables, he could express the common property as an equation relating these variables. To give a simple example, the unit circle is the locus of all points (in the plane) that are one unit distant from the center. If we choose the center at the origin of the coordinate system and use the Pythagorean Theorem, we get the equation of the unit circle: $x^2 + y^2 = 1$ (as already noted, this is a special case of the general quadratic equation). It should be noted that Descartes's coordinate system was not rectangular but oblique, and that he considered only positive coordinates, that is, points in the first quadrant—a far cry from the common practice today.

La Géométrie had an enormous influence on subsequent generations of mathematicians; among them was the young Newton, who bought a Latin translation and studied it on his own while a student at Cambridge. Descartes's work brought to a close classical Greek geometry, the essence of which was geometric construction and proof. From then on, geometry became inseparable from algebra, and soon from the calculus as well.

Pierre de Fermat was the exact opposite of Descartes. Whereas the mercurial Descartes constantly changed locations, allegiances, and careers, Fermat was a model of stability; indeed, so uneventful was his life that few stories about him exist. He began his career as a public servant and in 1631 became member of the *parlement* (court of law) of the city of Toulouse, a post he retained for the rest of his life. In his free time he studied languages, philosophy, literature, and poetry; but his main devotion was to mathematics, which he regarded as a kind of intellectual recreation. Whereas many of the mathematicians of his time were also physicists or astronomers, Fermat was the embodiment of the pure mathematician. His main interest was number theory, the "purest" of all branches of mathematics. Among his many contributions to this field is the assertion that the equation $x^n + y^n = z^n$ has no solutions in positive integers except when $n = 1$ and 2. The case $n = 2$ had already been known to the Greeks in connection with the Pythagorean Theorem. They knew that certain right triangles have sides whose lengths have integer values, such as the triangles with sides 3, 4, 5 or 5, 12, 13 (indeed, $3^2 + 4^2 = 5^2$ and $5^2 + 12^2 = 13^2$). So it was only natural to ask whether a similar equation for higher powers of x, y, and z could have integer solutions (we exclude the trivial cases 0, 0, 0 and 1, 0, 1). Fermat's answer was no. In the margin of his copy of Diophantus' *Arithmetica*, a classic work on number theory written in Alexandria in the third century A.D. and translated into Latin in 1621, he wrote: "To divide a cube into two other cubes, a fourth power, or in general, any power whatever into two powers of the same denomination above the second is impossible; I have found an admirable proof of this, but the margin is too narrow to contain it." Despite numerous attempts and many false claims, and thousands of special values of n for which the assertion has been shown to be true, the general statement remains unproved. Known as Fermat's Last Theorem ("theorem" is of course a misnomer), it is the most celebrated unsolved problem in mathematics.[1]

Closer to our subject, Fermat was interested in the quadrature of curves whose general equation is $y = x^n$, where n is a positive integer. These curves are sometimes called generalized parabolas (the parabola itself is the case $n = 2$). Fermat approximated the area under each curve by a series of rectangles whose bases form a decreasing geometric progression. This, of course, was very similar to Archimedes' method of exhaustion; but unlike his predecessor, Fermat did not shy away from summing up an infinite series. Figure 19 shows a portion of the curve $y = x^n$ between the points $x = 0$ and $x = a$ on the x-axis. We imagine that the interval from $x = 0$ to $x = a$ is divided into an infinite number of subintervals by the points . . . K, L, M, N, where

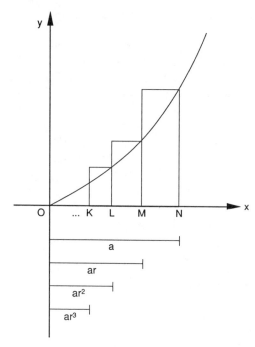

FIG. 19. Fermat's method of approximating the area under the graph of $y = x^n$ by a series of rectangles whose bases form a geometric progression.

$ON = a$. Then, starting at N and working backward, if these intervals are to form a decreasing geometric progression, we have $ON = a$, $OM = ar$, $OL = ar^2$, and so on, where r is less than 1. The heights (ordinates) to the curve at these points are then a^n, $(ar)^n$, $(ar^2)^n$, ... From this it is easy to find the area of each rectangle and then sum up the areas, using the summation formula for an infinite geometric series. The result is the formula

$$A_r = \frac{a^{n+1}(1 - r)}{1 - r^{n+1}}, \tag{1}$$

where the subscript r under the A indicates that this area still depends on our choice of r.[2]

Fermat then reasoned that in order to improve the fit between the rectangles and the actual curve, the width of each rectangle must be made small (fig. 20). To achieve this, the common ratio r must be close to 1—the closer, the better the fit. Alas, when $r \to 1$, equation 1 becomes the indeterminate expression 0/0. Fermat was able to get around this difficulty by noticing that the denominator of equation 1, $1 - r^{n+1}$, can be written in factored form as $(1 - r)(1 + r + r^2 + \ldots + r^n)$. When the factor $1 - r$ in the numerator and denominator is canceled, equation 1 becomes

$$A_r = \frac{a^{n+1}}{1 + r + r^2 + \ldots + r^n}.$$

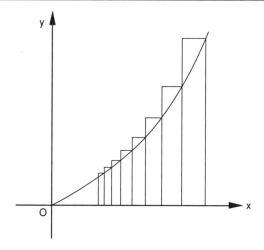

Fig. 20. A better approximation can be achieved by making the rectangles smaller while increasing their number.

As we let $r \to 1$, each term in the denominator tends to 1, resulting in the formula

$$A = \frac{a^{n+1}}{n+1}. \tag{2}$$

Every student of calculus will recognize equation 2 as the integration formula $\int_0^a x^n dx = a^{n+1}/(n+1)$. We should remember, however, that Fermat's work was done around 1640, some thirty years before Newton and Leibniz established this formula as part of their integral calculus.[3]

Fermat's work was a significant breakthrough because it accomplished the quadrature not just of one curve but of an entire family of curves, those given by the equation $y = x^n$ for positive integral values of n. (As a check, we note that for $n = 2$ the formula gives $A = a^3/3$, in agreement with Archimedes' result for the parabola.) Further, by slightly modifying his procedure, Fermat showed that equation 2 remains valid even when n is a *negative* integer, provided we now take the area from $x = a$ (where $a > 0$) to infinity.[4] When n is a negative integer, say $n = -m$ (where m is positive), we get the family of curves $y = x^{-m} = 1/x^m$, often called generalized hyperbolas. That Fermat's formula works even in this case is rather remarkable, since the equations $y = x^m$ and $y = x^{-m}$, despite their seeming similarity, represent quite different types of curves: the former are everywhere continuous, whereas the latter become infinite at $x = 0$ and consequently have a "break" (a vertical asymptote) there. We can well imagine Fermat's delight at discovering that his previous result remained valid even when the restriction under which it was originally obtained ($n = $ a positive integer) was removed.[5]

Alas, there was one snag. Fermat's formula failed for the one curve from which the entire family derives its name: the hyperbola $y =$

$1/x = x^{-1}$. This is because for $n = -1$, the denominator $n + 1$ in equation 2 becomes 0. Fermat's frustration at not being able to account for this important case must have been great, but he concealed it behind the simple words, "I say that all these infinite hyperbolas except the one of Appolonius [the hyperbola $y = 1/x$], or the first, may be squared by the method of geometric progression according to a uniform and general procedure."[6]

It remained for one of Fermat's lesser known contemporaries to solve the unyielding exceptional case. Grégoire (or Gregorius) de Saint-Vincent (1584–1667), a Belgian Jesuit, spent much of his professional life working on various quadrature problems, particularly that of the circle, for which he become known to his colleagues as a circle-squarer (it turned out that his quadrature in this case was false). His main work, *Opus geometricum quadraturae circuli et sectionum coni* (1647), was compiled from the thousands of scientific papers Saint-Vincent left behind when he hurriedly fled Prague before the advancing Swedes in 1631; these were rescued by a colleague and returned to their author ten years later. The delay in publication makes it difficult to establish Saint-Vincent's priority with absolute certainty, but it does appear that he was the first to notice that when $n = -1$, the rectangles used in approximating the area under the hyperbola all have *equal areas*. Indeed (see fig. 21), the widths of the successive rectangles, starting at N, are $a - ar = a(1 - r)$, $ar - ar^2 = ar(1 - r)$, . . . , and the heights at N, M, L, . . . are $a^{-1} = 1/a$, $(ar)^{-1} = 1/ar$, $(ar^2)^{-1} = 1/ar^2$, . . . ; the areas are therefore $a(1 - r) \cdot 1/a = 1 - r$,

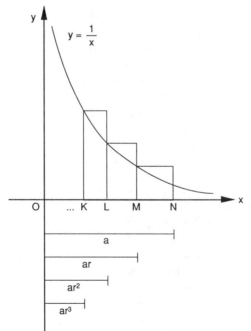

FIG. 21. Fermat's method applied to the hyperbola. Saint-Vincent noticed that when the bases form a geometric progression, the rectangles have equal areas; thus the area is proportional to the logarithm of the horizontal distance.

$ar(1 - r) \cdot 1/ar = 1 - r$, and so on. This means that as the distance from 0 grows geometrically, the corresponding areas grow in equal increments—that is, *arithmetically*—and this remains true even when we go to the limit as $r \to 1$ (that is, when we make the transition from the discrete rectangles to the continuous hyperbola). But this in turn implies that the relation between area and distance is *logarithmic*. More precisely, if we denote by $A(t)$ the area under the hyperbola from some fixed reference point $x > 0$ (for convenience we usually choose $x = 1$) to a variable point $x = t$, we have $A(t) = \log t$. One of Saint-Vincent's students, Alfonso Anton de Sarasa (1618–1667), wrote down this relation explicitly,[7] marking one of the first times that use was made of the logarithmic *function*, whereas until then logarithms were regarded mainly as a computational device.[8]

Thus the quadrature of the hyperbola was finally accomplished,

---

of Religion. 155

Problem $\dfrac{n}{n-1} - 1 = a$, which gives $n = \dfrac{a+1}{a}$;

fo the Equation to the Hyperbola fought, is

$$\overline{y\,x}\Big|^{\frac{a\times 1}{a}} = 1.$$

Let (as before) AC, AH be the Afymtotes of any Hyperbola DL F defined by this Equation $y\,x^n = 1$, in which the Abfciffa $AK = x$, and Ordinate $KL = y$, and n is fuppofed either equal to, or greater than Unity. 1°. It appears that in all Hyperbola's the interminate Space $CAKLD$ is infinite, and the interminate Space $HAGLF$ (except in the *Apollonian* where $n = 1$) is finite. 2°. In every Hyperbola, one Part of it continually approaches nearer and nearer to the Afymptote AC, and the other part continually nearer to the other Afymptote AH; that is, LD meets with AC at a Point infinitely diftant from A, and LF meets with AH at a Point infinitely diftant from A.

3°. In two different Hyperbola's DLF, dlf, if we fuppofe n to be greater in the Equation of dlf, than it is in the Equation of DLF, then LD fhall meet fooner with AC than

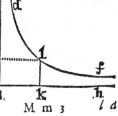

FIG. 22. A page from George Cheyene's *Philosophical Principles of Religion* (London, 1734), discussing the quadrature of the hyperbola.

some two thousand years after the Greeks had first tackled the problem. One question, however, still remained open: the formula $A(t) = \log t$ does indeed give the area under the hyperbola as a function of the variable t, but it is not yet suitable for numerical computations because no particular base is implied. To make the formula practicable, we must decide on a base. Would any base do? No, because the hyperbola $y = 1/x$ and the area under it (say from $x = 1$ to $x = t$) exist independently of any particular choice of a base. (The situation is analogous to the circle: we know that the general relationship between area and radius is $A = kr^2$, but we are not free to choose the value of k arbitrarily.) So there must be some particular "natural" base that determines this area numerically. As we shall see in Chapter 10, that base is the number e.

By the mid 1600s, then, the main ideas behind the integral calculus were fairly well known to the mathematical community.[9] The method of indivisibles, though resting on shaky grounds, had been successfully applied to a host of curves and solids; and Archimedes' method of exhaustion, in its revised modern form, solved the quadrature of the family of curves $y = x^n$. But successful as these methods were, they were not yet fused into a single, unified system; every problem required a different approach, and success depended on geometric ingenuity, algebraic skills, and a good deal of luck. What was needed was a general and systematic procedure—a set of algorithms—that would allow one to solve these problems with ease and efficiency. That procedure was provided by Newton and Leibniz.

Notes and Sources

1. As this book was going to press, it was announced that Andrew Wiles of Princeton University had finally proved the theorem (*New York Times*, 24 June 1993). His two-hundred-page proof, as yet unpublished, must still undergo careful scrutiny before the problem can be considered solved.

2. See Ronald Calinger, ed., *Classics of Mathematics* (Oak Park, Ill.: Moore Publishing Company, 1982), pp. 336–338.

3. John Wallis, whom we have already met in connection with his infinite product, independently arrived at the same result at about the same time as Fermat. The same formula for positive integers n was already known to several earlier mathematicians, among them Bonaventura Cavalieri (ca. 1598–1647), Gilles Persone de Roberval (1602–1675), and Evangelista Torricelli (1608–1647)—all pioneers in the method of indivisibles. On this subject see D. J. Struik, ed., *A Source Book in Mathematics, 1200–1800* (Cambridge, Mass.: Harvard University Press, 1969), ch. 4.

4. Actually, for $n = -m$ equation 2 gives the area with a negative sign; this is because the function $y = x^n$ is increasing when $n > 0$ and decreasing when $n < 0$ as one moves from left to right. The negative sign, however, is of no consequence as long as we consider the area in absolute value (just as we do with distance).

5. Both Fermat and Wallis later extended equation 2 to the case where n is a fraction p/q.

6. Calinger, ed., *Classics of Mathematics*, p. 337.

7. Margaret E. Baron, *The Origins of the Infinitesimal Calculus* (1969; rpt. New York: Dover, 1987), p. 147.

8. On the history of the hyperbolic area and its relation to logarithms, see Julian Lowell Coolidge, *The Mathematics of Great Amateurs* (1949; rpt. New York: Dover, 1963), pp. 141–146.

9. The origins of the differential calculus will be discussed in the next chapter.

The Birth of a New Science

*[Newton's] peculiar gift was the power of holding
conitnuously in his mind a purely mental problem until
he had seen through it.*
—JOHN MAYNARD KEYNES

Isaac Newton was born in Woolsthorpe in Lincolnshire, England, on Christmas Day (by the Julian calendar) 1642, the year of Galileo's death. There is a symbolism in this coincidence, for half a century earlier Galileo had laid the foundations of mechanics on which Newton would erect his grand mathematical description of the universe. Never has the biblical verse, "One generation passeth away, and another generation cometh: but the earth abideth for ever" (Ecclesiastes 1:4), been more prophetically fulfilled.[1]

Newton's early childhood was marked by family misfortunes. His father died a few months before Isaac was born; his mother soon remarried, only to lose her second husband too. Young Newton was thus left in the custody of his grandmother. At the age of thirteen he was sent to grammar school, where he studied Greek and Latin but very little mathematics. In 1661 he enrolled as a student at Trinity College, Cambridge University, and his life would never be the same.

As a freshman he studied the traditional curriculum of those days, with heavy emphasis on languages, history, and religion. We do not know exactly when or how his mathematical interests were sparked. He studied on his own the mathematical classics available to him: Euclid's *Elements*, Descartes's *La Géométrie*, Wallis's *Arithmetica infinitorum*, and the works of Viète and Kepler. None of these is easy reading even today, when most of the facts contained in them are well known; certainly they were not in Newton's time, when mathematical literacy was the privilege of a very few. The fact that he studied these works on his own, with no outside help and few friends with whom he could share his thoughts, set the stage for his future character as a reclusive genius who needed little outside inspiration to make his great discoveries.[2]

In 1665, when Newton was twenty-three years old, an outbreak of plague closed the Cambridge colleges. For most students this would have meant an interruption in their regular studies, possibly even ruining their future careers. The exact opposite happened to Newton. He returned to his home in Lincolnshire and enjoyed two years of complete freedom to think and shape his own ideas about the universe. These "prime years" (in his own words) were the most fruitful of his life, and they would change the course of science.[3]

Newton's first major mathematical discovery involved infinite series. As we saw in Chapter 4, the expansion of $(a + b)^n$ when n is a positive integer consists of the sum of $n + 1$ terms whose coefficients can be found from Pascal's triangle. In the winter of 1664/65 Newton extended the expansion to the case where n is a fraction, and in the following fall to the case where n is negative. For these cases, however, the expansion involves infinitely many terms—it becomes an *infinite series*. To see this, let us write Pascal's triangle in a form slightly different from the one we used earlier.

$n = 0$:	1	0	0	0	0	0	...
$n = 1$:	1	1	0	0	0	0	...
$n = 2$:	1	2	1	0	0	0	...
$n = 3$:	1	3	3	1	0	0	...
$n = 4$:	1	4	6	4	1	0	...

(This "staircase" version of the triangle first appeared in 1544 in Michael Stifel's *Arithmetica integra*, a work already mentioned in Chapter 1.) As we recall, the sum of the jth entry and the $(j - 1)$th entry in any row gives us the jth entry in the row below it, forming the pattern ⌐↓. The zeros at the end of each row simply indicate that the expansion is finite. To deal with the case where n is a negative integer, Newton extended the table *backward* (upward in our table) by computing the *difference* between the jth entry in each row and the $(j - 1)$th entry in the row *above* it, forming the pattern ↗. Knowing that each row begins with 1, he obtained the following array:

$n = -4$:	1	-4	10	-20	35	-56	84	...
$n = -3$:	1	-3	6	-10	15	-21	28	...
$n = -2$:	1	-2	3	-4	5	-6	7	...
$n = -1$:	1	-1	1	-1	1	-1	1	...
$n = 0$:	1	0	0	0	0	0	0	...
$n = 1$:	1	1	0	0	0	0	0	...
$n = 2$:	1	2	1	0	0	0	0	...
$n = 3$:	1	3	3	1	0	0	0	...
$n = 4$:	1	4	6	4	1	0	0	...

As an example, the 84 in the row for $n = -4$ is the difference between the 28 below it and the -56 to its left: $28 - (-56) = 84$. One consequence of this backward extension is that when n is negative, the expansion never terminates; instead of a finite sum, we get an infinite series.

To deal with the case where n is a fraction, Newton carefully studied the numerical pattern in Pascal's triangle until he was able to "read between the lines," to interpolate the coefficients when $n = 1/2$, $3/2$, $5/2$, and so on. For example, for $n = 1/2$ he got the coefficients 1, 1/2, –1/8, 1/16, –5/128, 7/256 . . . .[4] Hence the expansion of $(1 + x)^{1/2}$—that is, of $\sqrt{(1 + x)}$—is given by the infinite series $1 + (1/2)x - (1/8)x^2 + (1/16)x^3 - (5/128)x^4 + (7/256)x^5 - + \ldots$

Newton did not *prove* his generalization of the binomial expansion for negative and fractional n's; he merely conjectured it. As a double check, he multiplied the series for $(1 + x)^{1/2}$ term by term by itself and found, to his delight, that the result was $1 + x$.[5] And he had another clue that he was on the right track. For $n = -1$, the coefficients in Pascal's triangle are 1, –1, 1, –1, ... If we use these coefficients to expand the expression $(1 + x)^{-1}$ in powers of x, we get the infinite series

$$1 - x + x^2 - x^3 + - \ldots.$$

Now this is simply an infinite geometric series with the initial term 1 and common ratio $-x$. Elementary algebra teaches us that, *provided the common ratio is between –1 and 1*, the series converges precisely to $1/(1 + x)$. So Newton knew that his conjecture must be correct at least for this case. At the same time, it gave him a warning that one cannot treat infinite series in the same way as finite sums, because here the question of convergence is crucial. He did not use the word *convergence*—the concepts of limit and convergence were not yet known—but he was clearly aware that in order for his results to be valid, x must be sufficiently small.

Newton now formulated his binomial expansion in the following form:

$$(P + PQ)^{\frac{m}{n}} = P^{\frac{m}{n}} + \frac{m}{n} \cdot AQ + \frac{m-n}{2n} \cdot BQ + \frac{m-2n}{3n} \cdot CQ + \ldots$$

where A denotes the first term of the expansion (that is, $P^{\frac{m}{n}}$), B the second term, and so on (this is equivalent to the formula given in Chapter 4). Although Newton possessed this formula by 1665, he did not enunciate it until 1676 in a letter to Henry Oldenburg, secretary of the Royal Society, in response to Leibniz's request for more information on the subject. His reluctance to publish his discoveries was Newton's hallmark throughout his life, and it would bring about his bitter priority dispute with Leibniz.

Newton now used his binomial theorem to express the equations of various curves as infinite series in the variable x, or, as we would say

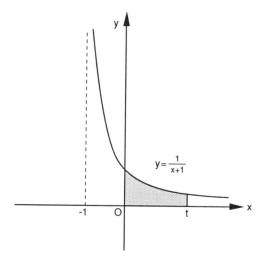

FIG. 23. The area under the hyperbola $y = 1/(x + 1)$ from $x = 0$ to $x = t$ is given by $\log (t + 1)$.

today, power series in x. He regarded these series simply as polynomials, treating them according to the ordinary rules of algebra. (Today we know that these rules do not always apply to infinite series, but Newton was unaware of these potential difficulties.) By applying Fermat's formula $x^{n+1}/(n + 1)$ to each term of the series (in modern language, term by term integration), he was able to effect the quadrature of many new curves.

Of particular interest to Newton was the equation $(x + 1)y = 1$, whose graph is the hyperbola shown in figure 23 (it is identical to the graph of $xy = 1$ but translated one unit to the left). If we write this equation as $y = 1/(x + 1) = (1 + x)^{-1}$ and expand it in powers of x, we get, as we have already seen, the series $1 - x + x^2 - x^3 + - \ldots$. Newton was aware of Saint-Vincent's discovery that the area bounded by the hyperbola $y = 1/x$, the x-axis, and the ordinates $x = 1$ and $x = t$ is $\log t$. This means that the area bounded by the hyperbola $y = 1/(x + 1)$, the x-axis, and the ordinates $x = 0$ and $x = t$ is $\log (t + 1)$ (see fig. 23). Thus, by applying Fermat's formula to each term of the equation

$$(1 + x)^{-1} = 1 - x + x^2 - x^3 + - \ldots$$

and considering the result as an equality between areas, Newton found the remarkable series

$$\log (1 + t) = t - \frac{t^2}{2} + \frac{t^3}{3} - \frac{t^4}{4} + - \ldots.$$

This series converges for all values of t in the interval $-1 < t \leq 1$ and in theory could be used to compute the logarithms of various numbers, although its slow rate of convergence makes such computations impractical.[6] Typically, Newton did not publish his discovery, and this time he had a good reason. In 1668 Nicolaus Mercator (ca. 1620–1687),[7] who was born in Holstein (then Denmark) and spent most of his years in England, published a work entitled *Logarithmotechnia* in which this series appeared for the first time (it was also discovered independently by Saint-Vincent). When Newton learned of Mercator's publication, he was bitterly disappointed, feeling that he had been deprived of the credit due to him. One would think that the incident should have prompted him to hasten publication of his discoveries in the future, but just the opposite happened. From then on, he would confide his work only to a close circle of friends and colleagues.

There was yet another player in the discovery of the logarithmic series. In the same year that Mercator published his work, William Brouncker (ca. 1620–1684), a founder of the Royal Society and its first president, showed that the area bounded by the hyperbola $(x + 1)y = 1$, the x-axis, and the ordinates $x = 0$ and $x = 1$ is given by the infinite series $1 - 1/2 + 1/3 - 1/4 + - \ldots$, or alternately by the series $1/(1 \cdot 2) + 1/(3 \cdot 4) + 1/(5 \cdot 6) + \ldots$ (the latter series can be obtained from the former by adding the terms in pairs). This result is the special case of Mercator's series for $t = 1$. Brouncker actually summed up sufficiently many terms of the series to arrive at the value 0.69314709, which he recognized as being "proportional" to log 2. We now know that the proportionality is actually an equality, because the logarithm involved in the quadrature of the hyperbola is the natural logarithm, that is, logarithm to the base e.

The confusion over who first discovered the logarithmic series is typical of the period just before the invention of the calculus, when many mathematicians were working independently on similar ideas and arriving at the same results. Many of these discoveries were never officially announced in a book or journal but were circulated as pamphlets or in personal correspondence to a small group of colleagues and students. Newton himself announced many of his discoveries in this way, a practice that was to have unfortunate consequences for him and for the scientific community at large. Fortunately, no serious priority dispute arose in the case of the logarithmic series, for Newton's mind was already set on a discovery of much greater consequence: the calculus.

The name "calculus" is short for "differential and integral calculus," which together constitute the two main branches of the subject (it is also known as the infinitesimal calculus). The word *calculus* itself has nothing to do with this particular branch of mathematics; in its broad sense it means any systematic manipulation of mathematical objects, whether numbers or abstract symbols. The Latin word *calculus* means a pebble, and its association with mathematics comes from the use of pebbles for reckoning—a primitve version of the familiar abacus. (The etymological root of the word is *calc* or *calx*, meaning lime, from which the words calcium and chalk are also derived.) The restricted meaning of the word *calculus*—that is, the differential and integral calculus—is due to Leibniz. Newton never used the word, preferring instead to call his invention the "method of fluxions."

The differential calculus is the study of change, and more specifically the *rate of change*, of a variable quantity. Most of the physical phenomena around us involve quantities that change with time, such as the speed of a moving car, the temperature readings of a thermometer, or the electric current flowing in a circuit. Today we call such a quantity a variable; Newton used the term *fluent*. The differential calculus is concerned with finding the rate of change of a variable, or, to use Newton's expression, the *fluxion* of a given fluent. His choice of words reveals his mind at work. Newton was as much a physicist as a mathematician. His worldview was a dynamic one, where everything was in a continual state of motion caused by the action of known forces. This view, of course, did not originate with Newton; attempts to explain all motion by the action of forces go back to antiquity and reached their climax with Galileo laying the foundations of mechanics in the early 1600s. But it was Newton who unified the host of known observational facts into one grand theory, his universal law of gravitation, which he enunciated in his *Philosophiae naturalis principia mathematica*, first published in 1687. His invention of the calculus, though not directly related to his work in physics (he rarely used it in the *Prinicpia* and was careful to cast his reasoning in geometric form when he did[8]), was no doubt influenced by his dynamic view of the universe.

Newton's point of departure was to consider two variables related to each other by an equation, say $y = x^2$ (today we call such a relation a *function*, and to indicate that y is a function of x we write $y = f(x)$). Such a relation is represented by a graph in the xy plane, in our example a parabola. Newton thought of the graph of a function as a curve generated by a moving point $P(x, y)$. As P traces the curve, both the x and the y coordinates continuously vary with time; time itself was thought to "flow" at a uniform rate—hence the word *fluent*. Newton now set out to find the rates of change of x and y with respect to time, that is, their fluxions. This he did by considering the difference, or change, in the values of x and y between two "adjacent" instances and then dividing this difference by the elapsed time interval. The final, crucial step was to set the elapsed time interval equal to 0—or, more precisely, to think of it as so small as to be negligible.

Let us see how this works for the function $y = x^2$. We consider a small time interval ε (Newton actually used the letter O, but because of its similarity to zero we will use ε). During this time interval, the x coordinate changes by the amount $\dot{x}\varepsilon$, where $\dot{x}$ is Newton's notation for the rate of change, or fluxion, of x (this became known as the "dot notation"). The change in y is likewise $\dot{y}\varepsilon$. Substituting $x + \dot{x}\varepsilon$ for x and $y + \dot{y}\varepsilon$ for y in the equation $y = x^2$, we get $y + \dot{y}\varepsilon = (x + \dot{x}\varepsilon)^2 = x^2 + 2x(\dot{x}\varepsilon) + (\dot{x}\varepsilon)^2$. But since $y = x^2$, we can cancel y on the left side of the equation with x^2 on the right side and obtain $\dot{y}\varepsilon = 2x(\dot{x}\varepsilon) + (\dot{x}\varepsilon)^2$. Dividing both sides by ε, we get $\dot{y} = 2x\dot{x} + \dot{x}^2\varepsilon$. The final step is to let ε be equal to 0, leaving us with $\dot{y} = 2x\dot{x}$. This is the relation between the fluxions of the two fluents x and y or, in modern language, between the rates of change of the variables x and y, each regarded as a function of time.

Newton gave several examples of how his "method of fluxion" works. The method is entirely general: it can be applied to any two fluents related to each other by an equation. By following the procedure as shown above, one obtains a relation between the fluxions, or rates of change, of the original variables. As an exercise, the reader may work out one of Newton's own examples, that of the cubic equation $x^3 - ax^2 + axy - y^3 = 0$. The resulting equation relating the fluxions of x and y is

$$3x^2\dot{x} - 2ax\dot{x} + ax\dot{y} + ay\dot{x} - 3y^2\dot{y} = 0.$$

This equation is more complicated than that for the parabola, but it serves the same purpose: it enables us to express the rate of change of x in terms of the rate of change of y and vice versa, for every point $P(x, y)$ on the curve.

But there is more to the method of fluxions than just finding the rates of change of the variables with respect to time. If we divide the fluxion of y by that of x (that is, compute the ratio $\dot{y}/\dot{x}$), we get the rate of change of y *with respect to x* . Now, this last quantity has a simple geometric meaning: it measures the steepness of the curve at each point on it. More precisely, the ratio $\dot{y}/\dot{x}$ is the *slope of the tangent line to the curve at the point P(x, y)*, where by *slope* we mean the rise-to-run ratio at that point. For example, for the parabola $y = x^2$ we found that the relation between the two fluxions is $\dot{y} = 2x\dot{x}$, so that $\dot{y}/\dot{x} = 2x$. This means that at each point $P(x, y)$ on the parabola, the tangent line has a slope equal to twice the value of the x coordinate at that point. If $x = 3$, the slope, or rise-to-run ratio, is 6; if $x = -3$, the slope is -6 (a negative slope means that the curve is going down as we move from left to right); if $x = 0$, the slope is 0 (this means that the parabola has a horizontal tangent line at $x = 0$); and so on (see fig. 24).

Let us emphasize the last point. Although Newton thought of x and y as varying with time, he ended up with a purely geometric interpretation of fluxions, one that does not depend on time. He needed the notion of time only as a mental aid to crystallize his ideas.

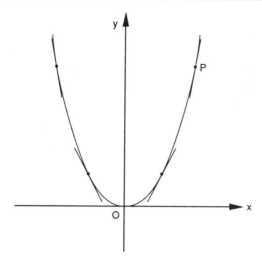

Fig. 24. Tangent lines to the parabola $y = x^2$.

Newton now applied his method to numerous curves, finding their slopes, their highest and lowest points (points of maximum and minimum), their curvature (the rate at which the curve changes direction), and their points of inflection (points where the curve changes from concave up to concave down or vice versa)—all geometric properties related to the tangent line. Because of its association with the tangent line, the process of finding the fluxion of a given fluent was known in Newton's time as the *tangent problem*. Today we call this process *differentiation*, and the fluxion of a function is called its *derivative*. Newton's dot notation has not survived either; except in physics, where it still appears occasionally, we use today Leibniz's much more efficient differential notation, as we shall see in the next chapter.

Newton's method of fluxions was not an entirely new idea. Just as with integration, it had been in the air for some time, and both Fermat and Descartes used it in several particular cases. The importance of Newton's invention was that it provided a *general procedure*—an algorithm—for finding the rate of change of practically any function. Most of the rules of differentiation that are now part of the standard calculus course were found by him; for example, if $y = x^n$, then $\dot{y} = nx^{n-1}\dot{x}$ (where n can have any value, positive or negative, integral or fractional, or even irrational). His predecessors had paved the way, but it was Newton who transformed their ideas into a powerful and universal tool, soon to be applied with enormous success to almost every branch of science.

Newton next considered the *inverse* of the tangent problem: given the fluxion, find the fluent. Generally speaking, this is a more difficult problem, just as division is a more difficult operation than multiplica-

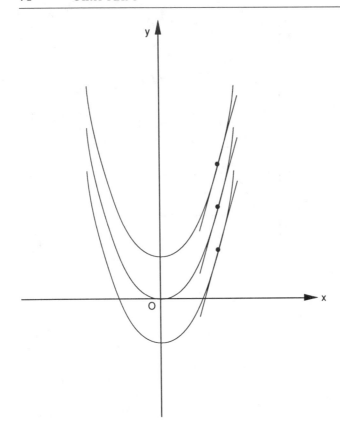

FIG. 25. The slope of the tangent line remains unchanged when the curve is moved up or down.

tion, or square-root extraction than squaring. In simple cases the result can be found by "guessing," as the following example shows. Given the fluxion $\dot{y} = 2x\dot{x}$, find the fluent y. An obvious answer is $y = x^2$, but $y = x^2 + 5$ would also be an answer, as would $x^2 - 8$ or in fact $x^2 + c$, where c is any constant. The reason is that the graphs of all these functions are obtained from the graph of $y = x^2$ by shifting it up or down, and hence they have the same slope at a given value of x (fig. 25). Thus a given fluxion has infinitely many fluents corresponding to it, differing from one another by arbitrary constants.

Having shown that the fluxion of $y = x^n$ is $\dot{y} = nx^{n-1}\dot{x}$, Newton next reversed the formula so that it now reads: If the fluxion is $\dot{y} = x^n\dot{x}$, then the fluent (apart from the additive constant) is $y = x^{n+1}/(n + 1)$. (We can check this result by differentiating, getting $\dot{y} = x^n\dot{x}$.) This formula, too, applies to fractional as well as integral values of n; to give one of Newton's own examples, if $\dot{y} = x^{1/2}\dot{x}$, then $y = (2/3)x^{3/2}$. But the formula fails for $n = -1$, for then the denominator becomes 0. This is the case where the fluxion is proportional to $1/x$, the very same case that had defied Fermat in his attempts to square the hyperbola. Newton knew (and we will shortly see how) that the result in this case involves logarithms; he called them "hyperbolic logarithms" to distinguish them from Briggs's "common" logarithms.

Today the process of finding the fluent of a given fluxion is called *indefinite integration*, or *antidifferentiation*, and the result of integrating a given function is its indefinite integral, or antiderivative (the "indefinite" refers to the existence of the arbitrary constant of integration). But Newton did more than just provide rules for differentiation and integration. We recall Fermat's discovery that the area under the curve $y = x^n$ from $x = 0$ to some $x > 0$ is given by the expression $x^{n+1}/(n + 1)$—the same expression that appears in the antidifferentiation of $y = x^n$. Newton recognized that this connection between area and antidifferentiation is not a coincidence; he realized, in other words, that the two fundamental problems of the calculus, the tangent problem and the area problem, are *inverse* problems. This is the crux of the differential and integral calculus.

Given a function $y = f(x)$, we can define a new function, $A(t)$, which represents the area under the graph of $f(x)$ from a given fixed value of x, say $x = a$, to some variable value $x = t$ (fig. 26). We will call this new function the *area function* of the original function. It is a function of t because if we change the value of t—that is, move the point $x = t$ to the right or left—the area under the graph will also change. What Newton realized amounts to this: *The rate of change of the area function with respect to t is equal, at every point $x = t$, to the value of the original function at that point*. Stated in modern terms, the derivative of $A(t)$ is equal to $f(t)$. But this in turn means that $A(t)$ itself is an *antiderivative* of $f(t)$. Thus, in order to find the area under the graph of $y = f(x)$, we must find an antiderivative of $f(x)$ (where we have replaced the variable t by x). It is in this sense that the two processes—finding the area and finding the derivative—are inverses

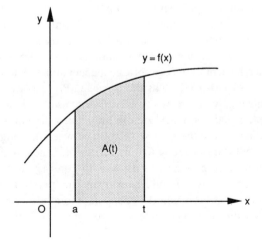

FIG. 26. The area under the graph of $y = f(x)$ from $x = a$ to $x = t$ is itself a function of t, denoted by $A(t)$.

of each other. Today this inverse relation is known as the Fundamental Theorem of the Calculus. As with the binomial theorem, Newton did not give a formal proof of the Fundamental Theorem, but he fully grasped its essence. Newton's discovery in effect merged the two branches of the calculus—until then regarded as separate, unrelated subjects—into a single unified field. (An outline of the proof of the Fundamental Theorem is found in Appendix 3.)

Let us illustrate this with an example. Suppose we wish to find the area under the parabola $y = x^2$ from $x = 1$ to $x = 2$. We first need to find an antiderivative of $y = x^2$; we already know that the antiderivatives of x^2 (note the use of the plural here) are given by $y = x^3/3 + c$, so that our area function is $A(x) = x^3/3 + c$. To determine the value of c, we note that at $x = 1$ the area must be 0, because this is the initial point of our interval; thus $0 = A(1) = 1^3/3 + c = 1/3 + c$, so that $c = -1/3$. Putting this value back in the equation for $A(x)$, we have $A(x) = x^3/3 - 1/3$. Finally, putting $x = 2$ in this last equation, we find $A(2) = 2^3/3 - 1/3 = 8/3 - 1/3 = 7/3$, the required area. If we consider how much labor was required to arrive at such a result using the method of exhaustion, or even the method of indivisibles, we can appreciate the enormous advantage of the integral calculus.

The invention of the calculus was the single most important event in mathematics since Euclid's compilation of the body of classical geometry in his *Elements* two thousand years earlier. It would forever change the way mathematicians think and work, and its powerful methods would affect nearly every branch of science, pure or applied. Yet Newton, who had a lifelong aversion to involvement in controversy (he had already been stung by the criticism of his views on the nature of light), did not publish his invention. He merely communicated it informally to his students and close colleagues at Cambridge. In 1669 he wrote a monograph, *De analysi per aequationes numero terminorum infinitas* (Of analysis by equations of an infinite number of terms), which he sent to his Cambridge teacher and colleague Isaac Barrow. Barrow (1630–1677) was the Lucasian Professor of Mathematics at Cambridge when Newton arrived there as a student, and his lectures on optics and geometry greatly influenced the young scientist. (Barrow knew about the inverse relation between the tangent and area problems but did not recognize its full significance, mainly because he used strictly geometric methods, in contrast to Newton's analytic approach.) Barrow would later resign his prestigious chair, ostensibly so that Newton could occupy it, though a more likely reason was his aspiration to become involved in the college's administrative and political life (which as occupant of the chair he was forbidden to do). Encouraged by Barrow, Newton in 1671 wrote an improved version of his invention, *De methodis serierum et fluxionum* (On the method of series and fluxions). A summary of this important work was not published until 1704, and then only as an appendix to Newton's major work, *Opticks* (the practice of annexing to a book an appendix on a subject unrelated to the main topic was quite common at the time). But it was not until 1736, nine years after Newton's death at the age of eighty-five, that the first full exposition of the subject was published as a book.

Thus for more than half a century the most important development in modern mathematics was known in England only to a small group of scholars and students centered around Cambridge. On the Continent, knowledge of the calculus—and the ability to use it—were at first confined to Leibniz and the two Bernoulli brothers.[9] Thus when Leibniz, one of Europe's leading mathematicians and philosophers, published his own version of the calculus in 1684, few mathematicians on the Continent had any doubt that his invention was indeed original. It was only some twenty years later that questions arose as to whether Leibniz had borrowed some of his ideas from Newton. The full consequences of Newton's procrastinations now became evident. The priority dispute about to erupt sent shock waves that would reverberate throughout the scientific community for the next two hundred years.

NOTES AND SOURCES

1. Every aspect of the life and work of this most famous mathematician of the modern era has been thoroughly researched and documented. For this reason, no specific source references will be given in this chapter to Newton's mathematical discoveries. Among the many works on Newton, perhaps the most authoritative are Richard S. Westfall, *Never at Rest: A Biography of Isaac Newton* (Cambridge: Cambridge University Press, 1980), which contains an extensive bibliographical essay, and *The Mathematical Papers of Isaac Newton*, ed. D. T. Whiteside, 8 vols. (Cambridge: Cambridge University Press, 1967–84).

2. We are reminded of another recluse of more recent times, Albert Einstein. Later in life both Newton and Einstein became prominent public figures, involving themselves in political and social affairs as their scientific output diminished. At the age of fifty-four Newton was offered, and accepted, the office of Warden of the Royal Mint, and at sixty-one he was elected president of the Royal Society, a position he retained for the rest of his life. At the age of seventy-three Einstein was offered the presidency of the State of Israel, an honor he turned down.

3. We are again reminded of Einstein, who shaped his special theory of relativity while enjoying the seclusion of his modest job at the Swiss Patent Office in Bern.

4. These coefficients can be written as 1, 1/2, $-1/(2 \cdot 4)$, $(1 \cdot 3)/(2 \cdot 4 \cdot 6)$, $-(1 \cdot 3 \cdot 5)/(2 \cdot 4 \cdot 6 \cdot 8)$, ...

5. Actually, Newton used the series for $(1 - x^2)^{1/2}$, which can be obtained from the series for $(1 + x)^{1/2}$ by formally replacing x by $-x^2$ in each term. His interest in this particular series stemmed from the fact that the function $y = (1 - x^2)^{1/2}$ describes the upper half of the unit circle $x^2 + y^2 = 1$. The series was already known to Wallis.

6. However, a variant of this series, $\log (1 + x)/(1 - x) = 2(x + x^3/3 + x^5/5 + \dots)$ for $-1 < x < 1$, converges much faster.

7. He is unrelated to the Flemish cartographer Gerhardus Mercator (1512–1594), inventor of the famous map projection named after him.

8. For the reasons, see W. W. Rouse Ball, *A Short Account of the History of Mathematics* (1908; rpt. New York: Dover, 1960), pp. 336–337.

9. Ibid., pp. 369–370. We are reminded again of Einstein, whose general theory of relativity was said to have been understood by only ten scientists when it was published in 1916.

9

The Great Controversy

If we must confine ourselves to one system of notation then
there can be no doubt that that which was invented by
Leibnitz is better fitted for most of the purposes to which the
infinitesimal calculus is applied than that of fluxions,
and for some (such as the calculus of variations) it is
indeed almost essential.
—W. W. ROUSE BALL, *A Short Account of the History of*
Mathematics (1908)

Newton and Leibniz will always be mentioned together as the co-inventors of the calculus. In character, however, the two men could hardly be less alike. Baron Gottfried Wilhelm von Leibniz (or Leibnitz) was born in Leipzig on 1 July 1646. The son of a philosophy professor, the young Leibniz soon showed great intellectual curiosity. His interests, in addition to mathematics, covered a wide range of topics, among them languages, literature, law, and above all, philosophy. (Newton's interests outside mathematics and physics were theology and alchemy, subjects on which he spent at least as much time as on his more familiar scientific work.) Unlike the reclusive Newton, Leibniz was a sociable man who loved to mix with people and enjoy the pleasures of life. He never married, which is perhaps the only trait he shared with Newton—apart, of course, from their interest in mathematics.

Among Leibniz's contributions to mathematics we should mention, in addition to the calculus, his work in combinatorics, his recognition of the binary system of numeration (the system that uses only two digits, 0 and 1, the basis of today's computers), and his invention of a calculating machine that could add as well as multiply (Pascal, some thirty years earlier, had built a machine that could only add). As a philosopher, he believed in a rational world in which everything follows reason and harmony. He attempted to develop a formal system of logic in which all deductions could be made in an algorithmic, computational manner. This idea was taken up almost two centuries later by the English mathematician George Boole (1815–1864), who founded what is now known as symbolic logic. We can see a common thread, a preoccupation with formal symbolism, running through these diverse interests. In mathematics, a good choice of symbols—a system of notation—is almost as important as the subject they represent, and the calculus is no exception. Leibniz's adeptness in formal symbolism would give his calculus an edge over Newton's method of fluxions, as we shall see.

Leibniz made his early career in law and diplomacy. The elector of Mainz employed him in both capacities and sent him abroad on various missions. In 1670, with Germany gripped by fear of an invasion by Louis XIV of France, Leibniz the diplomat came up with a strange idea: divert France's attention from Europe by letting it take Egypt, from where it could attack the Dutch possessions in southeast Asia. This plan did not win his master's approval, but more than a century later a similar scheme was indeed carried out when Napoleon Bonaparte invaded Egypt.

Notwithstanding the tense relations with France, Leibniz went to Paris in 1672 and for the next four years absorbed all the amenities, social as well as intellectual, that this beautiful city could offer. There he met Christian Huygens (1629–1695), Europe's leading mathematical physicist, who encouraged Leibniz to study geometry. Then in January 1673 he was sent on a dipolmatic mission to London, where he met several of Newton's colleagues, among them Henry Oldenburg (ca. 1618–1677), secretary of the Royal Society, and the mathematician John Collins (1625–1683). During a second brief visit in 1676, Collins showed Leibniz a copy of Newton's *De analysi*, which he had obtained from Isaac Barrow (see p. 80). This last visit would later become the focal point of the priority dispute between Newton and Leibniz.

Leibniz first conceived his differential and integral calculus around 1675, and by 1677 he had a fully developed and workable system. From the start, his approach differed from Newton's. As we have seen, Newton's ideas were rooted in physics; he regarded the fluxion as a rate of change, or velocity, of a moving point whose continuous motion generated the curve $y = f(x)$. Leibniz, who was much closer to philosophy than to physics, shaped his ideas in a more abstract way. He thought in terms of *differentials*: small increments in the values of the variables x and y.

Figure 27 shows the graph of a function $y = f(x)$ and a point $P(x, y)$ on it. We draw the tangent line to the graph at P and on it consider a neighboring point T. This gives us the small triangle PRT, which Leibniz called the *characteristic triangle*; its sides PR and RT are the increments in the x and y coordinates as we move from P to T. Leibniz denoted these increments by dx and dy, respectively. He now ar-

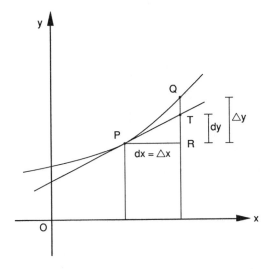

FIG. 27. Leibniz's characteristic triangle *PRT*. The ratio *RT/PR*, or *dy/dx*, is the slope of the tangent line to the curve at *P*.

gued that if *dx* and *dy* are sufficiently small, the tangent line to the graph at *P* will be almost identical to the graph itself in the neighborhood of *P*; more precisely, the line segment *PT* will very nearly coincide with the *curved* segment *PQ*, where *Q* is the point on the graph directly above or below *T*. To find the slope of the tangent line at *P*, we only need to find the rise-to-run ratio of the characteristic triangle, that is, the ratio *dy/dx*. Leibniz now reasoned that since *dx* and *dy* are small quantities (sometimes he thought of them as infinitely small), their ratio represents not only the slope of the tangent line at *P* but also the steepness of the *graph* at *P*. The ratio *dy/dx*, then, was Leibniz's equivalent of Newton's fluxion, or rate of change, of the curve.

There is one fundamental flaw in this argument. The tangent line, though very nearly identical with the curve near *P*, does not *coincide* with it. The two would coincide only if points *P* and *T* coincide, that is, when the characteristic triangle shrinks to a point. But then the sides *dx* and *dy* both become 0, and their ratio becomes the indeterminate expression 0/0. Today we get around this difficulty by defining the slope as a *limit*. Referring again to figure 27, we choose two neighboring points *P* and *Q*, both *on the graph*, and denote the sides *PR* and *RQ* of the triangle-like shape *PRQ* (really a curved shape) by Δx and Δy, respectively. (Note that Δx is equal to *dx*, but Δy is slightly different from *dy*; in fig. 27 Δy is larger than *dy* because *Q* is above *T*.) Now the rise-to-run ratio of the graph between *P* and *Q* is $\Delta y/\Delta x$. If we let both Δx and Δy approach 0, their ratio will approach a certain limiting value, and it is this limit that we denote today by *dy/dx*. In symbols, $dy/dx = \lim_{\Delta x \to 0}(\Delta y/\Delta x)$.

Let us summarize. What Leibniz denoted by *dy/dx* and thought of

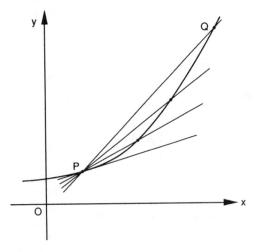

FIG. 28. As point Q moves toward point P, the secant lines PQ approach the tangent line at P.

as a ratio of two small increments is today written as $\Delta y/\Delta x$. Geometrically, the ratio $\Delta y/\Delta x$—called the *difference quotient*—is the slope of the *secant* line between P and Q (see fig. 28). As Δx approaches 0, the point Q moves back toward P along the graph, causing the secant line to turn slightly until, in the limit, it coincides with the tangent line.[1] It is the slope of the latter that we denote by dy/dx; it is called the *derivative of y with respect to x*.[2]

We see, then, that the limit concept is indispensable for defining the slope, or rate of change, of a function. But in Leibniz's time the limit concept was not yet known; the distinction between the ratio of two finite quantities, however small, and the *limit* of this ratio as the two quantities tend to 0, caused much confusion and raised serious questions about the very foundations of the differential calculus. These questions were not fully settled until the nineteenth century, when the limit concept was put on firm grounds.

To illustrate how Leibniz's idea works, let us find the derivative of the function $y = x^2$, using modern notation. If x is increased by an amount Δx, the corresponding increase in y is $\Delta y = (x + \Delta x)^2 - x^2$, which, after expanding and simplifying, becomes $2x\Delta x + (\Delta x)^2$. The difference quotient $\Delta y/\Delta x$ is therefore equal to $[2x\Delta x + (\Delta x)^2]/\Delta x = 2x + \Delta x$. If we let Δx tend to 0, $\Delta y/\Delta x$ will tend to $2x$, and it is this last expression that we denote by dy/dx. This result can be generalized: if $y = x^n$ (where n can be any number), then $dy/dx = nx^{n-1}$. This is identical with the result Newton obtained using his method of fluxions.

Leibniz's next step was to derive general rules for operating with the derivative dy/dx for various combinations of functions. Today these are known as the rules of differentiation, and they form the core of the standard calculus course. Here we summarize these rules using modern notation.

1. The derivative of a constant is 0. This is clear from the fact that the graph of a constant function is a horizontal straight line whose slope is everywhere 0.

2. If a function is multiplied by a constant, we need to differentiate only the function itself and then multiply the result by the constant. In symbols, if $y = ku$, where $u = f(x)$, then $dy/dx = k(du/dx)$. For example, if $y = 3x^2$, then $dy/dx = 3 \cdot (2x) = 6x$.

3. If y is the sum of two functions $u = f(x)$ and $v = g(x)$, its derivative is the sum of the derivatives of the individual functions. In symbols, if $y = u + v$, then $dy/dx = du/dx + dv/dx$. For example, if $y = x^2 + x^3$, then $dy/dx = 2x + 3x^2$. A similar rule holds for the difference of two functions.

4. If y is the product of two functions, $y = uv$, then $dy/dx = u(dv/dx) + v(du/dx)$.[3] For example, if $y = x^3(5x^2 - 1)$, then $dy/dx = x^3 \cdot (10x) + (5x^2 - 1) \cdot (3x^2) = 25x^4 - 3x^2$ (we could, of course, obtain the same result by writing $y = 5x^5 - x^3$ and differentiating each term separately). A slightly more complicated rule holds for the ratio of two functions.

5. Suppose that y is a function of the variable x and that x itself is a function of another variable t (time, for example); in symbols, $y = f(x)$ and $x = g(t)$. This means that y is an indirect function, or a *composite function*, of t: $y = f(x) = f[g(t)]$. Now, the derivative of y with respect to t can be found by multiplying the derivatives of the two component functions: $dy/dt = (dy/dx) \cdot (dx/dt)$. This is the famous "chain rule." On the surface it appears to be nothing more than the familiar cancelation rule of fractions, but we must remember that the "ratios" dy/dx and dx/dt are really *limits* of ratios, obtained by letting the numerator and denominator in each tend to 0. The chain rule shows the great utility of Leibniz's notation: we can manipulate the symbol dy/dx *as if* it were an actual ratio of two quantities. Newton's fluxional notation does not have the same suggestive power.

To illustrate the use of the chain rule, suppose that $y = x^2$ and $x = 3t + 5$. To find dy/dt, we simply find the "component" derivatives dy/dx and dx/dt and multiply them. We have $dy/dx = 2x$ and $dx/dt = 3$, so that $dy/dt = (2x) \cdot 3 = 6x = 6(3t + 5) = 18t + 30$. Of course, we could have arrived at the same result by substituting the expression $x = 3t + 5$ into y, expanding the result, and then differentiating it term by term: $y = x^2 = (3t + 5)^2 = 9t^2 + 30t + 25$, so that $dy/dt = 18t + 30$. In this example the two methods are about equally long; but if instead of $y = x^2$ we had, say, $y = x^5$, a direct computation of dy/dt would be quite lengthy, whereas applying the chain rule would be just as simple as for $y = x^2$.

Let us illustrate how these rules can be used to solve a practical problem. A ship leaves port at noon, heading due west at ten miles per hour. A lighthouse is located five miles north of the port. At 1 P.M., at what rate will the ship be receding from the lighthouse? Denoting

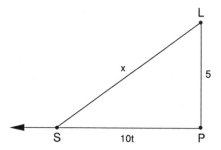

FIG. 29. One of numerous problems that can be solved easily with the aid of calculus: to find the rate at which a ship S, traveling in a given direction at a given speed, recedes from the lighthouse L.

the distance from the lighthouse to the ship at time t by x (fig. 29), we have by the Pythagorean Theorem $x^2 = (10t)^2 + 5^2 = 100t^2 + 25$, so that $x = \sqrt{(100t^2 + 25)} = (100t^2 + 25)^{1/2}$. This expression gives us the distance x as a function of the time t. To find the rate of change of x with respect to t, we regard x as a composition of two functions, $x = u^{1/2}$ and $u = 100t^2 + 25$. By the chain rule we have $dx/dt = (dx/du) \cdot (du/dt) = (1/2u^{-1/2}) \cdot (200t) = 100t \cdot (100t^2 + 25)^{-1/2} = 100t/\sqrt{(100t^2 + 25)}$. At 1 P.M. we have $t = 1$, giving us a rate of change of $100/\sqrt{125} \approx 8.944$ miles per hour.

The second part of the calculus is the integral calculus, and here again Leibniz's notation proved superior to Newton's. His symbol for the antiderivative of a function $y = f(x)$ is $\int y dx$, where the elongated S is called an (indefinite) *integral* (the dx merely indicates that the variable of integration is x). For example, $\int x^2 dx = x^3/3 + c$, as can be verified by differentiating the result. The additive constant c comes from the fact that any given function has infinitely many antiderivatives, obtained from one another by adding an arbitrary constant (see p. 78); hence the name "indefinite" integral.

Just as he had done with differentiation, Leibniz developed a set of formal rules for integration. For example, if $y = u + v$, where u and v are functions of x, then $\int y dx = \int u dx + \int v dx$, and similarly for $y = u - v$. These rules can be verified by differentiating the result, in much the same way that the result of a subtraction can be verified by addition. Unfortunately, there is no general rule for integrating a product of two functions, a fact that makes integration a much more difficult process than differentiation.

Leibniz's conception of integration differed from Newton's not only in notation. Whereas Newton viewed integration as the inverse of differentiation (given a fluxion, find the fluent), Leibniz started with the area problem: given a function $y = f(x)$, find the area under the graph of $f(x)$ from some fixed value of x, say $x = a$ to a variable value $x = t$. He thought of the area as the sum of many narrow strips

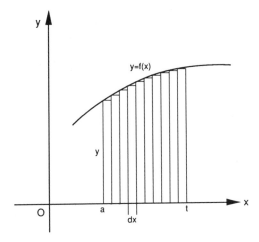

FIG. 30. Leibniz regarded the area under the graph of $y = f(x)$ as the sum of a large number of narrow rectangles, each with a base dx and a height $y = f(x)$.

of width dx and heights y that vary with x according to the equation $y = f(x)$ (fig. 30). By adding up the areas of these strips, he got the total area under the graph: $A = \int y\,dx$. His symbol $\int$ for integration is reminiscent of an elongated S (for "sum"), just as his differentiation symbol d stood for "difference."

As we saw earlier, the idea of finding the area of a given shape by regarding it as the sum of a large number of small shapes originated with the Greeks, and Fermat successfully used it to effect the quadrature of the family of curves $y = x^n$. But it is the Fundamental Theorem of Calculus—the inverse relation between differentiation and integration—that transformed the new calculus into such a powerful tool, and the credit for its formulation goes to Newton and Leibniz alone. As we saw in Chapter 8, the theorem involves the area under the graph of $f(x)$. Denoting this area by $A(x)$ (because it is itself a function of x),[4] the theorem says that the rate of change, or derivative, of $A(x)$ at every point x is equal to $f(x)$; in symbols, $dA/dx = f(x)$. But this in turn means that $A(x)$ is an *antiderivative* of $f(x)$: $A(x) = \int f(x)dx$. These two inverse relations are the crux of the entire differential and integral calculus. In abbreviated notation, we can write them as:

$$\frac{dA}{dx} = y \iff A = \int y\,dx.$$

Here y is short for $f(x)$, and the symbol $\iff$ ("if and only if") means that each statement implies the other (that is, the two statements are equivalent). Newton also arrived at the same result, but it was Leibniz's superior notation that expressed the inverse relation between differentiation and integration (that is, between the tangent and area problems) so clearly and concisely.

In Chapter 8 we demonstrated the use of the Fundamental Theorem to find the area under the graph of $y = x^2$ from $x = 1$ to $x = 2$ (p. 80). Let us repeat this example using Leibniz's notation and taking the area from $x = 0$ to $x = 1$. We have $A(x) = \int x^2 dx = x^3/3 + c$. Now $A(0) = 0$, since $x = 0$ is the initial point of our interval; thus $0 = 0^3/3 + c$ and hence $c = 0$. Our area function is therefore $A(x) = x^3/3$, and the required area is $A(1) = 1^3/3 = 1/3$. In modern notation we write this as $A = {}_0\int^1 x^2 dx = (x^3/3)_{x=1} - (x^3/3)_{x=0} = 1^3/3 - 0^3/3 = 1/3$.[5] Thus, almost effortlessly, we arrive at the same result that had demanded of Archimedes, using the method of exhaustion, such a great deal of ingenuity and labor (p. 43).[6]

Leibniz published his differential calculus in the October 1684 issue of *Acta eruditorum*, the first German science journal, which he and his colleague Otto Mencke had founded two years earlier. His integral calculus was published in the same journal two years later, although the term *integral* was not coined until 1690 (by Jakob Bernoulli, about whom we will have more to say later).

✧ ✧ ✧

As early as 1673 Leibniz had been corresponding with Newton through Henry Oldenburg. From this correspondence Leibniz got a glimpse—but only a glimpse—of Newton's method of fluxions. The secretive Newton only hinted vaguely that he had discovered a new method of finding tangents and quadratures of algebraic curves. In response to Leibniz's request for further details, Newton, after much prodding from Oldenburg and Collins, replied in a manner that was common at the time: he sent Leibniz an anagram—a coded message of garbled letters and numbers—that no one could possibly decode but that could later serve as "proof" that he was the discoverer:

6accdæ13eff7i3l9n4o4qrr4s8t12vx.

This famous anagram gives the number of different letters in the Latin sentence "Data æquatione quotcunque fluentes quantitates involvente, fluxiones invenire: et vice versa" (Given an equation involving any number of fluent quantities, to find the fluxions, and vice versa).

Newton sent the letter to Oldenburg in October 1676 with a request that its content be transmitted to Leibniz. Leibniz received it in the summer of 1677 and immediately replied, again through Oldenburg, with a full account of his own differential calculus. He expected Newton to respond with equal openness, but Newton, increasingly suspicious that his invention might be claimed by others, refused to continue the correspondence.

Nevertheless, relations between the two remained cordial; they respected each other's work, and Leibniz lavished praise on his colleague: "Taking mathematics from the beginning of the world to the time when Newton lived, what he had done was much the better half."[7] Even the publication of Leibniz's calculus in 1684 did not immediately affect their relationship. In the first edition of the *Principia* (1687), his great treatise on the principles of mechanics, Newton acknowledged Leibniz's contribution—but added that Leibniz's method "hardly differed from mine, except in his forms of words and symbols."

For the next twenty years their relations remained more or less unchanged. Then, in 1704, the first official publication of Newton's method of fluxions appeared in an appendix to his *Opticks*. In the preface to this appendix Newton mentioned his 1676 letter to Leibniz, adding that "some years ago I lent out a manuscript containing such theorems [about the calculus]; and having since met with some things copied out of it, I have on this occasion made it public." Newton was, of course, referring to Leibniz's second visit to London in 1676, at which time Collins showed him a copy of *De analysi*. This thinly veiled hint that Leibniz had copied his ideas from Newton did not go unnoticed by Leibniz. In an anonymous review of Newton's earlier tract on quadrature, published in *Acta eruditorum* in 1705, Leibniz reminded his readers that "the elements of this calculus have been given to the public by its inventor, Dr. Wilhelm Leibniz, in these *Acta*." While not denying that Newton invented his fluxional calculus independently, Leibniz pointed out that the two versions of the calculus differed only in notation, not in substance, implying that in fact it was Newton who had borrowed his ideas from Leibniz.

This was too much for Newton's friends, who now rallied to defend his reputation (he himself, at this stage, remained behind the scene). They openly accused Leibniz of taking his ideas from Newton. Their most effective ammunition was Collins's copy of *De analysi*. Although Newton discusses the fluxional calculus only briefly in this tract (most of it deals with infinite series), the fact that Leibniz not only saw it during his 1676 visit to London but also took extensive notes from it exposed him to charges that he had indeed used Newton's ideas in his own work.

Accusations were now hurled back and forth across the English Channel, and soon the exchange became acrimonious. More and more persons joined the battle, some with a genuine desire to defend the reputation of their respective mentors, others with an eye toward settling personal accounts. As might be expected, Newton received unanimous support in England, while continental Europe stood behind Leibniz. One of Leibniz's staunchest supporters was Johann Bernoulli, brother of Jakob. The two Bernoullis were instrumental in making Leibniz's calculus known throughout Europe. In a letter published in 1713 Johann questioned the personal character of Newton.

Although Bernoulli later retracted his charges, Newton was stung to reply to him personally: "I have never grasped at fame among foreign nations, but I am very desirous to preserve my character for honesty, which the author of that epistle, as if by the authority of a great judge, had endeavoured to wrest from me. Now that I am old, I have little pleasure in mathematical studies, and I have never tried to propagate my opinions over the world, but have rather taken care not to involve myself in disputes on account of them."[8]

Newton was not as modest as his words might suggest. True, he shied away from controversies, but he ruthlessly pursued his enemies. In 1712, in response to Leibniz's request that his name be cleared of accusations of plagiarism, the Royal Society took up the matter. That distinguished body of scholars, whose president at the time was none other than Newton, appointed a committee to investigate the dispute and settle it once and for all. The committee was composed entirely of Newton's supporters, including the astronomer Edmond Halley, who was also one of Newton's closest friends (it was Halley who, after relentless prodding, persuaded Newton to publish his *Principia*). Its final report, issued in the same year, sidestepped the issue of plagiarism but concluded that Newton's method of fluxions preceded Leibniz's differential calculus by fifteen years. Thus, under the semblance of academic objectivity, the issue was supposedly settled.

But it was not. The dispute continued to poison the atmosphere in academic circles long after the two protagonists had died. In 1721, six years after Leibniz's death, the eighty-year-old Newton supervised a second printing of the Royal Society's report, in which he made numerous changes intended to undermine Leibniz's credibility. But even that did not satisfy Newton's desire to settle the account. In 1726, one year before his own death, Newton saw the publication of the third and final edition of his *Principia*, from which he deleted all mention of Leibniz.

The two great rivals differed no less in death than in life. Leibniz, embittered by the long priority dispute, spent his last years in almost complete neglect. His mathematical creativity came to an end, though he still wrote on philosophical matters. His last employer, George Ludwig, elector of Hanover, assigned him the task of writing the history of the royal family. In 1714 the elector became King George I of England, and Leibniz hoped that he might be invited to join the king in England. By then, however, the elector had lost interest in Leibniz's services. Or perhaps he wished to avoid the embarrassment that Leibniz's presence would have created in England, where Newton's popularity was at its peak. Leibniz died in 1716 at the age of seventy, almost completely forgotten. Only his secretary attended his funeral.

Newton, as we have seen, spent his last years pursuing his dispute with Leibniz. But far from being forgotten, he became a national hero. The priority dispute only increased his reputation, for by then it was seen as a matter of defending the honor of England against "attacks" from the Continent. Newton died on 20 March 1727 at the age of eighty-five. He was given a state funeral and buried in Westminster Abbey in London with honors normally reserved for statesmen and generals.

✧ ✧ ✧

Knowledge of the calculus was at first confined to a very small group of mathematicians: Newton's circle in England, and Leibniz and the Bernoulli brothers on the Continent. The Bernoullis spread it throughout Europe by teaching it privately to several mathematicians. Among them was the Frenchman Guillaume François Antoine L'Hospital (1661–1704), who wrote the first textbook on the subject, *Analyse des infiniment petits* (1696).[9] Other continental mathematicians caught up, and soon the calculus became the dominant mathematical topic of the eighteenth century. It was quickly expanded to cover a host of related topics, notably differential equations and the calculus of variations. These subjects fall under the broad category of *analysis*, the branch of mathematics that deals with change, continuity, and infinite processes.

In England, where it originated, the calculus fared less well. Newton's towering figure discouraged British mathematicians from pursuing the subject with any vigor. Worse, by siding completely with Newton in the priority dispute, they cut themselves off from developments on the Continent. They stubbornly stuck to Newton's dot notation of fluxions, failing to see the advantages of Leibniz's differential notation. As a result, over the next hundred years, while mathematics flourished in Europe as never before, England did not produce a single first-rate mathematician. When the period of stagnation finally ended around 1830, it was not in analysis but in algebra that the new generation of English mathematicans made their greatest mark.

NOTES AND SOURCES

1. This argument supposes that the function is *continuous* at *P*—that its graph does not have a break there. At points of discontinuity a function does not have a derivative.

2. The name "derivative" comes from Joseph Louis Lagrange, who also introduced the symbol $f'(x)$ for the derivative of $f(x)$; see p. 95.

3. This follows from the fact that an increment of Δx in x causes u to

increase by Δu and v by Δv; hence y increases by $\Delta y = (u + \Delta u)(v + \Delta v) - uv = u\Delta v + v\Delta u + \Delta u\Delta v$. Since (to paraphrase Leibniz) Δu and Δv are small, their product $\Delta u\Delta v$ is even smaller in comparison to the other terms and can therefore be ignored. We thus get $\Delta y \approx u\Delta v + v\Delta u$, where $\approx$ means "approximately equal to." Dividing both sides of this relation by Δx and letting Δx tend to 0 (and consequently changing the Δ's into d's), we get the required result.

4. Strictly speaking, one must make a distinction between x as the independent variable of the function $y = f(x)$ and x as the variable of the area function $A(x)$. On p. 79 we made this distinction by denoting the latter by t; the Fundamental Theorem then says that $dA/dt = f(t)$. It is common practice, however, to use the same letter for both variables, so long as there is no danger of confusion. We have followed this practice here.

5. The symbol $_a\int^b f(x)dx$ is called the *definite integral* of $f(x)$ from $x = a$ to $x = b$, the adjective "definite" indicating that no arbitrary constant is involved. Indeed, if $F(x)$ is an antiderivative of $f(x)$, we have $_a\int^b f(x)dx = [F(x) + c]_{x=b} - [F(x) + c]_{x=a} = [F(b) + c] - [F(a) + c] = F(b) - F(a)$, so that the constant c cancels out.

6. Note that the result obtained here gives the area *under* the parabola $y = x^2$ between the x-axis and the ordinates $x = 0$ and $x = 1$, while Archimedes' result (p. 43) gives the area of the sector inscribed *inside* the parabola. A moment's thought will show that the two results are compatible.

7. Quoted in Forest Ray Moulton, *An Introduction to Astronomy* (New York: Macmillan, 1928), p. 234.

8. Quoted in W. W. Rouse Ball, *A Short Account of the History of Mathematics* (1908; rpt. New York: Dover, 1960), pp. 359–60.

9. See Julian Lowell Coolidge, *The Mathematics of Great Amateurs* (1949; rpt. New York: Dover, 1963), pp. 154–163, and D. J. Struik, ed., *A Source Book in Mathematics, 1200–1800* (Cambridge, Mass.: Harvard University Press, 1969), pp. 312–316.

The Evolution of a Notation

$\mathbf{A}$ working knowledge of a mathematical topic requires a good system of notation. When Newton invented his "method of fluxions," he placed a dot over the letter representing the quantity whose fluxion (derivative) he sought. This dot notation—Newton called it the "pricked letter" notation—is cumbersome. To find the derivative of $y = x^2$, one must first obtain a relation between the fluxions of x and y with respect to time (Newton thought of each variable as "flowing" uniformly with time, hence the term *fluxion*), in this case $\dot{y} = 2x\dot{x}$ (see p. 75). The derivative, or rate of change, of y with respect to x is the ratio of the two fluxions, namely $\dot{y}/\dot{x} = 2x$.

The dot notation survived in England for more than a century and can still be found in physics textbooks to denote differentiation with respect to time. Continental Europe, however, adopted Leibniz's more efficient differential notation, dy/dx. Leibniz thought of dx and dy as small increments in the variables x and y; their ratio gave him a measure of the rate of change of y with repsect to x. Today we use the letter Δ (Greek capital delta) to denote Leibniz's differentials. His dy/dx is written as $\Delta y/\Delta x$, whereas dy/dx denotes the *limit* of $\Delta y/\Delta x$ as Δx and Δy approach 0.

The notation dy/dx for the derivative enjoys many advantages. It is highly suggestive and in many ways behaves like an ordinary fraction. For example, if $y = f(x)$ and $x = g(t)$, then y is an indirect function of t, $y = h(t)$. To find the derivative of this *composite function*, we use the "chain rule": $dy/dt = (dy/dx) \cdot (dx/dt)$. Note that although each derivative is a *limit* of a ratio, it behaves as if it were an actual ratio of two finite quantities. Similarly, if $y = f(x)$ is a one-to-one function (see p. 175), it has an inverse, $x = f^{-1}(y)$. The derivative of this inverse function is the reciprocal of the original derivative: $dx/dy = 1/(dy/dx)$, a formula that again mimics the way ordinary fractions behave.

Yet another notation for the derivative has the advantage of brevity: if $y = f(x)$, we denote its derivative by $f'(x)$ or simply y'. Thus, if $y = x^2$, then $y' = 2x$. We can write this even shorter in a single statement: $(x^2)' = 2x$. This notation was published in 1797 by Joseph Louis Lagrange (1736–1813) in his treatise *Théorie des fonctiones analytiques*, in which he also proposed the notation fx for a function of x—the precursor of our familiar $f(x)$. He called $f'x$ the *derived function* of *fx*, from which the modern term *derivative* comes. For the second derivative of y (see p. 104) he wrote y'' or $f''x$, and so on.

If u is a function of two independent variables, $u = f(x, y)$, we must specify with respect to which variable, x or y, we are differentiating. For this purpose we use the German ∂ instead of the Roman d and get the two *partial derivatives* of u: $\partial u/\partial x$ and $\partial u/\partial y$. In this notation all variables except those indicated are kept constant. For example, if $u = 3x^2y^3$, then $\partial u/\partial x = 3(2x)y^3 = 6xy^3$ and $\partial u/\partial y = 3x^2(3y^2) = 9x^2y^2$, where in the first case y is held constant, and in the second case x.

Sometimes we wish to refer to an operation without actually performing it. Symbols such as $+$, $-$, and $\sqrt{}$ are called operational symbols, or simply *operators*. An operator acquires a meaning only when it is applied to a quantity on which it can operate; for example, $\sqrt{16} = 4$. To indicate differentiation, we use the operator symbol d/dx, with the understanding that everything appearing to the *right* of the operator is to be differentiated, whereas everything to the left is not. For example, $x^2 \, d/dx(x^2) = x^2 \cdot 2x = 2x^3$. A second differentiation is denoted by $d/dx(d/dx)$, abbreviated as $d^2/(dx^2)$.

Here again a shorter notation has been devised: the differential operator D. This operator acts on any function standing to its immediate right, whereas quantities on its left are unaffected; for example, $x^2Dx^2 = x^2 \cdot 2x = 2x^3$. For a second differentiation we write D^2; thus $D^2x^5 = D(Dx^5) = D(5x^4) = 5 \cdot 4x^3 = 20x^3$. Similarly, D^n (where n is any positive integer) indicates n successive differentiations. Moreover, by allowing n to be a *negative* integer, we can extend the symbol D to indicate antidifferentiation (that is, indefinite integration; see p. 79). For example, $D^{-1}x^2 = x^3/3 + c$, as can easily be verified by differentiating the right side (here c is an arbitrary constant).

Since the function $y = e^x$ is equal to its own derivative, we have the formula $Dy = y$. This formula, of course, is merely a differential equation whose solution is $y = e^x$, or more generaly $y = Ce^x$. However, it is tempting to regard the equation $Dy = y$ as an ordinary algebraic equation and "cancel" the y on both sides, as if the symbol D were an ordinary quantity multiplied by y. Succumbing to this temptation, we get $D = 1$, an operational equation that, by itself, has no meaning; it regains its meaning only if we "remultiply" both sides by y.

Still, this kind of formal manipulation makes the operator D useful in solving certain types of differential equations. For example, the differential equation $y'' + 5y' - 6y = 0$ (a linear equation with constant coefficients) can be written as $D^2y + 5Dy - 6y = 0$. Pretending that all the symbols in this equation are ordinary algebraic quantities, we can "factor out" the unknown function y on the left side and get $(D^2 + 5D - 6)y = 0$. Now, a product of two factors can equal 0 only if either one or the other of the factors is 0. So we have either $y = 0$ (this is a trivial solution, being of no interest), or $D^2 + 5D - 6 = 0$. Again acting as if D were an algebraic quantity, we can factor this last expression and get $(D - 1)(D + 6) = 0$. Equating each factor to 0, we get the "solutions" $D = 1$ and $D = -6$. Of course, these solutions are merely operational statements; we must still "multiply" them by y, getting $Dy = y$ and $Dy = -6y$. The first equation has the solution $y = e^x$, or, more generally $y = Ae^x$, where A is an arbitrary constant. The second equation has the solution $y = Be^{-6x}$, where B is another arbitrary constant. Since the original equation is linear and its right side is equal to 0, the sum of the two solutions, namely $y = Ae^x + Be^{-6x}$, is also a solution—in fact, it is the *general* solution of the equation $y'' + 5y' - 6y = 0$.

The symbol D as an operator was first used in 1800 by the Frenchman Louis François Antoine Arbogast (1759–1803), although Johann Bernoulli had used it earlier in a non-operational sense. It was the English electrical engineer Oliver Heaviside (1850–1925) who elevated the use of operational methods to an art in its own right. By cleverly manipulating the symbol D and treating it as an algebraic quantity, Heaviside solved numerous applied problems, particularly differential equations arising in electric theory, in an elegant and efficient way. Heaviside had no formal mathematical education, and his carefree virtuosity in manipulating D was frowned upon by professional mathematicians. He defended his methods by maintaining that the end justified the means: his methods produced correct results, so their rigorous justification was of secondary importance to him. Heavisde's ideas did find their proper formal justification in the more advanced method known as the Laplace transform.[1]

NOTE

1. See Murray R. Spiegel, *Applied Differential Equations*, 3d ed. (Englewood Cliffs, N.J.: Prentice-Hall, 1981), pp. 168–169 and 204–211. For a fuller account of the evolution of the differentiation notation, see Florian Cajori, *A History of Mathematical Notations*, vol. 2, *Higher Mathematics* (1929; rpt. La Salle, Ill.: Open Court, 1951), pp. 196–242.

10

e^x: The Function That Equals Its Own Derivative

The natural exponential function is identical with its derivative. This is the source of all the properties of the exponential function and the basic reason for its importance in applications.

—RICHARD COURANT AND HERBERT ROBBINS, *What Is Mathematics?* (1941)

When Newton and Leibniz developed their new calculus, they applied it primarily to *algebraic curves*, curves whose equations are polynomials or ratios of polynomials. (A *polynomial* is an expression of the form $a_n x^n + a_{n-1} x^{n-1} + \ldots + a_1 x + a_0$; the constants a_i are the *coefficients*, and n, the *degree* of the polynomial, is a non-negative integer. For example, $5x^3 + x^2 - 2x + 1$ is a polynomial of degree 3.) The simplicity of these equations, and the fact that many of them show up in applications (the parabola $y = x^2$ is a simple example), made them a natural choice for testing the new methods of the calculus. But in applications one also finds many curves that do not fall in the category of algebraic curves. These are the *transcendental* curves (the term was coined by Leibniz to imply that their equations go beyond those studied in elementary algebra). Foremost among them is the exponential curve.

We saw in Chapter 2 how Henry Briggs improved Napier's logarithmic tables by introducing the base 10 and working with powers of this base. In principle, any positive number other than 1 can be a base. If we denote the base by b and its exponent by x, we get the *exponential function base* b, $y = b^x$. Here x represents any real number, positive or negative. We must, however, clarify what we mean by b^x when x is not an integer. When x is a rational number m/n, we define b^x to be either $\sqrt[n]{b^m}$ or $(\sqrt[n]{b})^m$—the two expressions are equal provided m/n is reduced to lowest terms; for example, $8^{2/3} = \sqrt[3]{8^2} = \sqrt[3]{64} = 4$, or $8^{2/3} = (\sqrt[3]{8})^2 = 2^2 = 4$. But when x is *irrational*—when it cannot be written as a ratio of two integers—this definition is useless. In this case we approximate the value of x by a *sequence of rational numbers*, which, in the limit, converge to x. Take as an example $3^{\sqrt{2}}$. We can think of the exponent $x = \sqrt{2} = 1.414213\ldots$ (an irrational number) as the limit of an infinite sequence of terminating decimals $x_1 = 1, x_2 = 1.4, x_3 = 1.41, x_4 = 1.414, \ldots$, each of which is a rational number. Each of these x_i's determines a unique value of 3^{x_i}, so we define $3^{\sqrt{2}}$ as the limit of the sequence 3^{x_i} as $i \to \infty$. With a hand-held calculator we can easily find the first few values of this sequence: $3^1 = 3$, $3^{1.4} = 4.656$, $3^{1.41} = 4.707$, $3^{1.414} = 4.728$, and so on (all rounded to three decimal places). In the limit we get 4.729, the desired value.

There is, of course, a subtle but crucial assumption behind this idea: as the x_i's converge to the limit $\sqrt{2}$, the corresponding values of 3^{x_i} converge to the limit $3^{\sqrt{2}}$. In other words, we assume that the function $y = 3^x$—and more generally, $y = b^x$—is a *continuous* function of x, that it varies smoothly, with no breaks or jumps. The assumption of continuity is at the heart of the differential calculus. It is already implied in the definition of the derivative, for when we compute the limit of the ratio $\Delta y / \Delta x$ as $\Delta x \to 0$, we assume that Δx *and* Δy tend to 0 simultaneously.

To see the general features of the exponential function, let us choose the base 2. Confining ourselves to integral values of x, we get the following table:

x	−5	−4	−3	−2	−1	0	1	2	3	4	5
2^x	1/32	1/16	1/8	1/4	1/2	1	2	4	8	16	32

If we plot these values in a coordinate system, we get the graph shown in figure 31. We see that as x increases, so does y—slowly at first, then at an ever faster rate to infinity. And conversely, when x

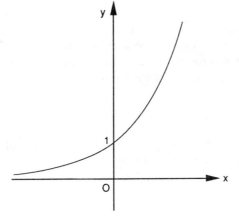

FIG. 31. The graph of an increasing exponential function.

decreases, y decreases at an ever slower rate; it will never reach 0, but come closer and closer to it. The negative x-axis is thus a horizontal asymptote of the function, the graphic equivalent of the limit concept discussed in Chapter 4.

The rate of growth of an exponential function can be quite astounding. A famous legend about the inventor of the game of chess has that, when summoned to the king and asked what reward he would wish for his invention, he humbly requested that one grain of wheat be put on the first square of the board, two grains on the second square, four grains on the third, and so on until all sixty-four squares were covered. The king, surprised by the modesty of this request, immediately ordered a sack of grain brought in, and his servants patiently began to place the grains on the board. To their astonishment, it soon became clear that not even all the grain in the kingdom would suffice to fulfill the request, for the number of grains on the last square, 2^{63}, is 9,223,372,036,854,775,808 (to which we must add the grains of all the previous squares, making the total number about twice as large). If we placed that many grains in an unbroken line, the line would be some two light-years long—about half the distance to the star Alpha Centauri, our closest celestial neighbor beyond the solar system.

The graph shown in figure 31 is typical of all exponential graphs, regardless of their base.[1] The simplicity of this graph is striking: it lacks most of the common features of the graphs of algebraic functions, such as x-intercepts (points where the graph crosses the x-axis), points of maximum and minimum, and inflection points. Furthermore, the graph has no *vertical* asymptotes—values of x near which the function increases or decreases without bound. Indeed, so simple is the exponential graph that we could almost dismiss it as uninteresting were it not for one feature that makes this graph unique: its rate of change.

As we saw in Chapter 9, the rate of change, or derivative, of a function $y = f(x)$ is defined as $dy/dx = \lim_{\Delta x \to 0} \Delta y / \Delta x$. Our goal is to find this rate of change for the function $y = b^x$. If we increase the value of x by Δx, y will increase by the amount $\Delta y = b^{x+\Delta x} - b^x$. Using the rules of exponentiation, we can write this as $b^x b^{\Delta x} - b^x$ or $b^x(b^{\Delta x} - 1)$. The required rate of change is thus

$$\frac{dy}{dx} = \lim_{\Delta x \to 0} \frac{b^x(b^{\Delta x} - 1)}{\Delta x}. \tag{1}$$

At this point it would be expedient to replace the symbol Δx by a single letter h, so that equation 1 becomes

$$\frac{dy}{dx} = \lim_{h \to 0} \frac{b^x(b^h - 1)}{h}. \tag{2}$$

We can make a second simplification by removing the factor b^x from the limit sign; this is because the limit in equation 2 involves only the variable h, whereas x is to be regarded as fixed. We thus arrive at the expression

$$\frac{dy}{dx} = b^x \lim_{h \to 0} \frac{b^h - 1}{h}.$$ (3)

Of course, at this point we have no guarantee that the limit appearing in equation 3 exists at all; the fact that it does exist is proved in advanced texts,[2] and we will accept it here. Denoting this limit by the letter k, we arrive at the following result:

$$\text{If } y = b^x, \text{ then } \frac{dy}{dx} = kb^x = ky.$$ (4)

This result is of such fundamental importance that we rephrase it in words: *The derivative of an exponential function is proportional to the function itself.*

Note that we have used the phrase "the derivative of *an* exponential function," not *the* exponential function, because until now the choice of b was entirely arbitrary. But the question now arises: Is there any particular value of b that would be especially convenient? Going back to equation 4, if we could choose b so as to make the proportionality constant k equal to 1, this clearly would make equation 4 particularly simple; it would, indeed, be the "natural" choice of b. Our task, then, is to determine the value of b for which k will be equal to 1, that is

$$\lim_{h \to 0} \frac{b^h - 1}{h} = 1.$$ (5)

It takes a bit of algebraic manipulation (and some subtle mathematical pedantry) to "solve" this equation for b, and we will omit the details here (a heuristic derivation is given in Appendix 4). The result is

$$b = \lim_{h \to 0} (1 + h)^{1/h}.$$ (6)

Now if in this equation we replace $1/h$ by the letter m, then as $h \to 0$, m will tend to infinity. We therefore have

$$b = \lim_{m \to \infty} (1 + 1/m)^m.$$ (7)

But the limit appearing in equation 7 is none other than the number $e = 2.71828 \ldots$ [3] We thus arrive at the following conclusion: *If the number e is chosen as base, the exponential function is equal to its own derivative.* In symbols,

If $y = e^x$, then $\dfrac{dy}{dx} = e^x$.   (8)

But there is more to this result. Not only is the function e^x equal to its own derivative, it is the *only* function (apart from a multiplicative constant) that has this property. To put it differently, if we solve the equation $dy/dx = y$ (a differential equation) for the function y, we get the solution $y = Ce^x$, where C is an arbitrary constant. This solution represents a family of exponential curves (fig. 32), each corresponding to a different value of C.

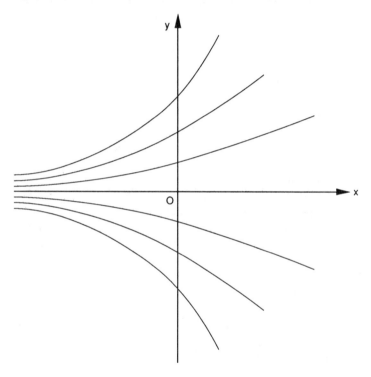

FIG. 32. The family of exponential curves $y = Ce^x$. Each graph corresponds to one value of C.

The central role of the function e^x—henceforth to be called the natural exponential function, or simply *the* exponential function—in mathematics and science is a direct consequence of these facts. In applications one finds numerous phenomena in which the rate of change of some quantity is proportional to the quantity itself. Any such phenomenon is governed by the differential equation $dy/dx = ay$, where the constant a determines the rate of change in each case. The solution is $y = Ce^{ax}$, where the arbitrary constant C is determined from the *initial condition* of the system: the value of y when $x = 0$.

Depending on whether a is positive or negative, y will increase or decrease with x, resulting in an *exponential growth* or *decay*. (When a is negative, one usually replaces it by $-a$, where a itself is now positive.) Let us note a few examples of such phenomena.

1. The rate of decay of a radioactive substance—and the amount of radiation it emits—is at every moment proportional to its mass m: $dm/dt = -am$. The solution of this differential equation is $m = m_0 e^{-at}$, where m_0 is the initial mass of the substance (the mass at $t = 0$). We see from this solution that m will gradually approach 0 but never reach it—the substance will never completely disintegrate. This explains why, years after nuclear material has been disposed as waste, it can still be a hazard. The value of a determines the rate of decay of the substance and is usually measured by the *half-life* time, the time it takes a radioactive substance to decay to one-half of its initial mass. Different substances have vastly different half-life times. For example, the common isotope of uranium (U^{238}) has a half-life of about five billion years, ordinary radium (Ra^{226}) about sixteen hundred years, while Ra^{220} has a half-life of only twenty-three milliseconds. This explains why some of the unstable elements in the periodic table are not found in natural minerals: whatever quantity may have been present when the earth was born has long since been transformed into more stable elements.

2. When a hot object at temperature T_0 is put in an environment of temperature T_1 (itself assumed to remain constant), the object cools at a rate proportional to the difference $T - T_1$ between its temperature at time t and the surrounding temperature: $dT/dt = -a(T - T_1)$. This is known as Newton's law of cooling. The solution is $T = T_1 + (T_0 - T_1)e^{-at}$, showing that T will gradually approach T_1 but never reach it.

3. When sound waves travel through air (or any other medium), their intensity is governed by the differential equation $dI/dx = -aI$, where x is the distance traveled. The solution, $I = I_0 e^{-ax}$, shows that the intensity decreases exponentially with distance. A similar law, known as Lambert's law, holds for the absorption of light in a transparent medium.

4. If money is compounded *continuously* (that is, every instant) at an annual interest rate r, the balance after t years is given by the formula $A = Pe^{rt}$, where P is the principal. Thus the balance grows exponentially with time.

5. The growth of a population follows an approximate exponential law.

The equation $dy/dx = ax$ is a *first-order* differential equation: it involves only the unknown function and its derivative. But most laws of physics are expressed in terms of *second-order* differential equations—equations involving the *rate of change of the rate of change* of a function, or its *second derivative*. For example, the acceleration of a moving object is the rate of change of its velocity; and since the velocity itself is the rate of change of distance, it follows that the acceleration is the rate of change of the rate of change, or the second derivative, of the distance. Since the laws of classical mechanics are based on Newton's three laws of motion—the second of which relates the acceleration of a body of mass m to the force acting on it ($F = ma$)—these laws are expressed in terms of second-order differential equations. A similar situation holds in electricity.

To find the second derivative of a function $f(x)$, we first differentiate $f(x)$ to get its first derivative; this derivative is itself a function of x, denoted by $f'(x)$. We then differentiate $f'(x)$ to obtain the second derivative, $f''(x)$. For example, if $f(x) = x^3$, then $f'(x) = 3x^2$ and $f''(x) = 6x$. There is, of course, nothing to stop us here; we can go on and find the third derivative, $f'''(x) = 6$, the fourth derivative (0), and so on. With a polynomial function of degree n, n successive differentiations will give us a constant, and all subsequent derivatives will be 0. For other types of functions, repeated differentiation may result in increasingly complex expressions. In applications, however, we rarely need to go beyond the second derivative.

Leibniz's notation for the second derivative is $d/dx(dy/dx)$, or (counting the d's as if they were algebraic quantities) $d^2y/(dx)^2$. Like the symbol dy/dx for the first derivative, this symbol, too, behaves in a way reminiscent of the familiar rules of algebra. For example, if we compute the second derivative of the product $y = u \cdot v$ of two functions $u(x)$ and $v(x)$, we get, after applying the product rule twice,

$$\frac{d^2y}{dx^2} = u\frac{d^2v}{dx^2} + 2\frac{du}{dx}\frac{dv}{dx} + v\frac{d^2u}{dx^2}.$$

This result, known as Leibniz's rule, bears a striking similarity to the binomial expansion $(a + b)^2 = a^2 + 2ab + b^2$. In fact, we can extend it to the *nth order* derivative of $u \cdot v$; the coefficients turn out to be exactly the binomial coefficients of the expansion of $(a + b)^n$ (see p. 32).

A frequent problem in mechanics is that of describing the motion of a vibrating system—a mass attached to a spring, for example—taking into account the resistance of the surrounding medium. This problem leads to a second-order differential equation with constant coefficients. An example of such an equation is

$$\frac{d^2y}{dt^2} + 5\frac{dy}{dt} + 6y = 0.$$

To solve this equation, let us make a clever guess: the solution is of the form $y = Ae^{mt}$, where A and m are as yet undetermined constants. Substituting this tentative solution in the differential equation, we get

$$e^{mt}(m^2 + 5m + 6) = 0,$$

which is an algebraic equation in the unknown m. Since e^{mt} is never 0, we can cancel it and get the equation $m^2 + 5m + 6 = 0$, known as the *characteristic equation* of the given differential equation (note that the two equations have the same coefficients). Factoring it, we get $(m + 2)(m + 3) = 0$, and after equating each factor to 0 we find the required values of m, namely -2 and -3. We thus have two distinct solutions, Ae^{-2t} and Be^{-3t}, and we can easily verify that their sum, $y = Ae^{-2t} + Be^{-3t}$, is also a solution—in fact, it is the *complete* solution of the differential equation. The constants A and B (which until now were arbitrary) can be found from the initial conditions of the system: the values of y and dy/dt when $t = 0$.

This method works with any differential equation of the kind just solved; to find the solution we need only to solve the characteristic equation. There is one snag, however: the characteristic equation may have *imaginary* solutions, solutions that involve the square root of -1. For example, the equation $d^2y/dx^2 + y = 0$ has the characteristic equation $m^2 + 1 = 0$, whose two solutions are the imaginary numbers $\sqrt{-1}$ and $-\sqrt{-1}$. If we denote these numbers by i and $-i$, the solution of the differential equation is $y = Ae^{ix} + Be^{-ix}$, where as before A and B are arbitrary constants.[4] But in all our encounters with the exponential function we have always assumed that the exponent is a real number. What, then, does an expression like e^{ix} mean? It was one of the great achievements of eighteenth-century mathematics that a meaning was given to the function e^{mx} even when m is imaginary, as we shall see in Chapter 13.

One other aspect of the exponential function must be considered. Most functions $y = f(x)$, when defined in an appropriate domain, have an inverse; that is, not only can we determine a unique value of y for every value of x in the domain, but we can also find a unique x for every permissible y. The rule that takes us back from y to x defines the *inverse function* of $f(x)$, denoted by $f^{-1}(x)$.[5] For example, the function $y = f(x) = x^2$ assigns to every real number x a unique $y \geq 0$, namely, the square of x. If we restrict the domain of $f(x)$ to non-negative numbers, we can reverse this process and assign to every $y \geq 0$ a unique x, the *square root* of y: $x = \sqrt{y}$.[6] It is customary to interchange the letters in this last equation so as to let x denote the independent variable and y the dependent variable; denoting the inverse function by f^{-1}, we thus get $y = f^{-1}(x) = \sqrt{x}$. The graphs of $f(x)$ and $f^{-1}(x)$ are mirror reflections of each other in the line $y = x$, as shown in figure 33.

Our goal is to find the inverse of the exponential function. We start with the equation $y = e^x$ and think of y as being given; we then wish to solve this equation for x, that is, express x in terms of y. We recall that the Briggsian or *common logarithm* of a number $y > 0$ is the

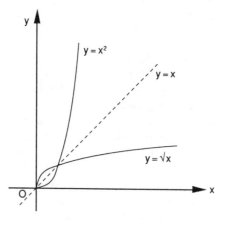

FIG. 33. The equations $y = x^2$ and $y = \sqrt{x}$ represent inverse functions; their graphs are mirror images of each other in the line $y = x$.

number x for which $10^x = y$. In exactly the same way, the *natural logarithm* of a number $y > 0$ is the number x for which $e^x = y$. And just as we write $x = \log y$ for the common logarithm (logarithm base 10) of y, so we write $x = \ln y$ for its natural logarithm (logarithm base e). The inverse of the exponential function, then, is the natural logarithmic function, and its equation, after interchanging x and y, is $y = \ln x$. Figure 34 shows the graphs of $y = e^x$ and and $y = \ln x$ plotted in the same coordinate system; as with any pair of inverse functions, the two graphs are mirror reflections of each other in the line $y = x$.

Having defined the natural logarithm as the inverse of the exponential function, we now wish to find its rate of change. Here again Leibniz's differential notation is of great help. It says that the rate of change of the inverse function is the *reciprocal* of (one divided

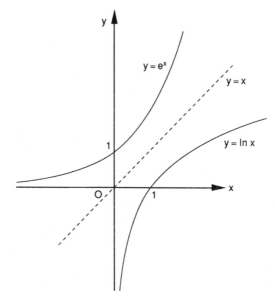

FIG. 34. The equations $y = e^x$ and $y = \ln x$ represent inverse functions.

by) the rate of change of the original function; in symbols, $dx/dy = 1/(dy/dx)$. For example, in the case of $y = x^2$ we have $dy/dx = 2x$, so that $dx/dy = 1/2x = 1/(2\sqrt{y})$. When we interchange x and y, our result reads: If $y = \sqrt{x}$, then $dy/dx = 1/(2\sqrt{x})$; even more briefly, $d(\sqrt{x})/dx = 1/(2\sqrt{x})$.

In the example just given, we could have found the same result by writing $y = \sqrt{x} = x^{1/2}$ and differentiating directly by using the power rule: $dy/dx = (1/2)x^{-1/2} = 1/(2\sqrt{x})$. But this is only because the inverse of a power function is again a power function, for which we know the rule of differentiation. In the case of the exponential function we must start from scratch. We have $y = e^x$ and $dy/dx = e^x = y$, so that $dx/dy = 1/e^x = 1/y$. This says that the rate of change of x—considered as a function of y—is equal to $1/y$. But what *is* x as a function of y? It is precisely $\ln y$, because $y = e^x$ is equivalent to $x = \ln y$. When we interchange letters as before, our formula reads: if $y = \ln x$, then $dy/dx = 1/x$; even more briefly, $d(\ln x)/dx = 1/x$. And this in turn means that $\ln x$ is an *antiderivative* of $1/x$: $\ln x = \int (1/x)dx$.[7]

We saw in Chapter 8 that the antiderivative of x^n is $x^{n+1}/(n + 1) + c$; in symbols, $\int x^n dx = x^{n+1}/(n + 1) + c$, where c is the constant of integration. This formula holds for all values of n except -1, since then the denominator $n + 1$ is 0. But when $n = -1$, the function whose antiderivative we are seeking is the hyperbola $y = x^{-1} = 1/x$—the same hyperbola whose quadrature Fermat had failed to carry out. The formula $\int (1/x)dx = \ln x + c$ now provides the "missing case." It explains at once Saint-Vincent's discovery that the area under the hyperbola follows a logarithmic law (p. 67). Denoting this area by $A(x)$, we have $A(x) = \ln x + c$. If we choose the initial point from which the area is reckoned as $x = 1$, we have $0 = A(1) = \ln 1 + c$. But $\ln 1 = 0$ (because $e^0 = 1$), so we have $c = 0$. We thus conclude: *The area under the hyperbola $y = 1/x$ from $x = 1$ to any $x > 1$ is equal to $\ln x$.*

Since the graph of $y = 1/x$ for $x > 0$ lies entirely above the x-axis, the area under it grows continuously the farther we move to the right; in mathematical language, the area is a *monotone increasing* function of x. But this means that as we start from $x = 1$ and move to the right, we will eventually reach a point x for which the area is exactly equal to 1. For this particular x we then have $\ln x = 1$, or (remembering the definition of $\ln x$), $x = e^1 = e$. This result at once gives the number e a geometric meaning that relates it to the hyperbola in much the same way as π is related to the circle. Using the letter A to denote area, we have:

Circle: $A = \pi r^2 \Rightarrow A = \pi$ when $r = 1$

Hyperbola: $A = \ln x \Rightarrow A = 1$ when $x = e$

Note, however, that the similarity is not perfect: whereas π is interpreted as the area of a unit circle, e is the *linear* dimension for which the area under the hyperbola is 1. Still, the analogous roles of the two most famous numbers in mathematics give us reason to suspect that perhaps there is an even deeper connection between them. And this is indeed the case, as we shall see in Chapter 13.

Notes and Sources

1. If the base is a number between 0 and 1, say 0.5, the graph is a mirror reversal of that shown in figure 31: it decreases from left to right and approaches the positive x-axis as $x \to \infty$. This is because the expression $y = 0.5^x = (1/2)^x$ can be written as 2^{-x}, whose graph is a mirror reflection of the graph of $y = 2^x$ in the y-axis.

2. See, for example, Edmund Landau, *Differential and Integral Calculus* (1934), trans. Melvin Hausner and Martin Davis (1950; rpt. New York: Chelsea Publishing Company, 1965), p. 41.

3. It is true that in Chapter 4 we defined e as the limit of $(1 + 1/n)^n$ for *integral* values of n, as $n \to \infty$. The same definition, however, holds even when n tends to infinity through all *real* values, that is, when n is a continuous variable. This follows from the fact that the function $f(x) = (1 + 1/x)^x$ is continuous for all $x > 0$.

4. If the characteristic equation has a *double* root m (that is, two equal roots), it can be shown that the solution of the differential equation is $y = (A + Bt)e^{mt}$. For example, the differential equation $d^2y/dt^2 - 4dy/dt + 4y = 0$, whose characteristic equation $m^2 - 4m + 4 = (m - 2)^2 = 0$ has the double root $m = 2$, has the solution $y = (A + Bt)e^{2t}$. For details, see any text on ordinary differential equations.

5. This symbol is somewhat unfortunate because it can easily be confused with $1/f(x)$.

6. The reason for restricting the domain of $y = x^2$ to $x \geq 0$ is to ensure that no two x values will give us the same y; otherwise the function would not have a unique inverse, since, for example $3^2 = (-3)^2 = 9$. In the terminology of algebra, the equation $y = x^2$ for $x \geq 0$ defines a one-to-one function.

7. This result gives rise to an alternative definition of the natural logarithmic function, as we show in Appendix 5.

The Parachutist

Among the numerous problems whose solution involves the exponential function, the following is particularly interesting. A parachutist jumps from a plane and at $t = 0$ opens his chute. At what speed will he reach the ground?

For relatively small velocities, we may assume that the resisting force exerted by the air is proportional to the speed of descent. Let us denote the proportionality constant by k and the mass of the parachutist by m. Two opposing forces are acting on the parachutist: his weight mg (where g is the acceleration of gravity, about 9.8 m/sec^2), and the air resistance kv (where $v = v(t)$ is the downward velocity at time t). The net force in the direction of motion is thus $F = mg - kv$, where the minus sign indicates that the force of resistance acts in a direction opposite to the direction of motion.

Newton's second law of motion says that $F = ma$, where $a = dv/dt$ is the acceleration, or rate of change of the velocity with respect to time. We thus have

$$m \frac{dv}{dt} = mg - kv. \tag{1}$$

Equation 1 is the *equation of motion* of the problem; it is a linear differential equation with $v = v(t)$ the unknown function. We can simplify equation 1 by dividing it by m and denoting the ratio k/m by a:

$$\frac{dv}{dt} = g - av \qquad (a = \frac{k}{m}). \tag{2}$$

If we consider the expression dv/dt as a ratio of two differentials, we can rewrite equation 2 so that the two variables v and t are separated, one on each side of the equation:

$$\frac{dv}{g - av} = dt. \tag{3}$$

We now integrate each side of equation 3—that is, find its antiderivative. This gives us

$$-\frac{1}{a} \ln (g - av) = t + c, \tag{4}$$

where ln stands for the natural logarithm (logarithm base e) and c is the constant of integration. We can determine c from the *initial condition*: the velocity at the instant the parachute opens. Denoting this velocity by v_0, we have $v = v_0$ when $t = 0$; substituting this into equation 4, we find $-1/a \ln (g - av_0) = 0 + c = c$. Putting this value of c back into equation 4, we get, after a slight simplification,

$$-\frac{1}{a}[\ln (g - av) - \ln (g - av_0)] = t.$$

But by the rules of logarithms we have $\ln x - \ln y = \ln x/y$, so we can write the last equation as

$$\ln \left[\frac{g - av}{g - av_0}\right] = -at. \tag{5}$$

Finally, solving equation 5 for v in terms of t, we get

$$v = \frac{g}{a}(1 - e^{-at}) + v_0 e^{-at}. \tag{6}$$

This is the required solution $v = v(t)$.

Two conclusions can be drawn from equation 6. First, if the parachutist opens his chute immediately upon jumping from the aircraft, we have $v_0 = 0$, so that the last term in equation (6) drops. But even if he falls freely before opening his chute, the effect of the initial velocity v_0 diminishes exponentially as time progresses; indeed, for $t \to \infty$, the expression e^{-at} tends to 0, and a *limiting velocity* $v_\infty = g/a = mg/k$ will be attained. This limiting velocity is independent of v_0; it depends only on the parachutist's weight mg and the resistance coefficient k. It is this fact that makes a safe landing possible. A graph of the function $v = v(t)$ is shown in figure 35.

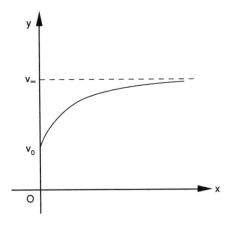

Fig. 35. A parachutist falling through the air attains a limiting velocity v_∞.

Can Perceptions Be Quantified?

In 1825 the German physiologist Ernst Heinrich Weber (1795–1878) formulated a mathematical law that was meant to measure the human response to various physical stimuli. Weber performed a series of experiments in which a blindfolded man holding a weight to which smaller weights were gradually added was asked to respond when he first felt the increase. Weber found that the response was proportional not to the absolute increase in weight but to the *relative* increase. That is, if the person could still feel an increase in weight from ten pounds to eleven pounds (a 10 percent increase), then, when the original weight was changed to twenty pounds, the corresponding threshold increase was two pounds (again a 10 percent increase); the threshold response to a forty-pound weight was four pounds, and so on. Expressed mathematically,

$$ds = k \frac{dW}{W}, \tag{1}$$

where ds is the threshold increase in response (the smallest increase still discernible), dW the corresponding increase in weight, W the weight already present, and k a proportionality constant.

Weber then generalized his law to include *any* kind of physiological sensation, such as the pain felt in response to physical pressure, the perception of brightness caused by a source of light, or the perception of loudness from a source of sound. Weber's law was later popularized by the German physicist Gustav Theodor Fechner (1801–1887) and became known as the Weber-Fechner law.

Mathematically, the Weber-Fechner law as expressed in equation 1 is a differential equation. Integrating it, we have

$$s = k \ln W + C, \tag{2}$$

where ln is the natural logarithm and C the integration constant. If we denote by W_0 the lowest level of physical stimulus that just barely causes a response (the threshold level), we have $s = 0$ when $W = W_0$, so that $C = -k \ln W_0$. Putting this back into equation 2 and noting that $\ln W - \ln W_0 = \ln W/W_0$, we finally get

$$s = k \ln \frac{W}{W_0}. \tag{3}$$

This shows that the response follows a logarithmic law. In other words, for the response to increase in equal steps, the corresponding stimulus must be increased in a constant *ratio*, that is, in a geometric progression.

Although the Weber-Fechner law seems to apply to a wide range of physiological responses, its universal validity has been a matter of contention. Whereas physical stimuli are objective quantities that can be precisely measured, the human response to them is a subjective matter. How do we measure the feeling of pain? Or the sensation of heat? There is one sensation, however, that *can* be measured with great precision: the sensation of musical pitch. The human ear is an extremely sensitive organ that can notice the change in pitch caused by a frequency change of only 0.3 percent. Professional musicians are acutely aware of the slightest deviation from the correct pitch, and even an untrained ear can easily tell when a note is off by a quarter tone or less.

When the Weber-Fechner law is applied to pitch, it says that equal musical intervals (increments in pitch) correspond to equal *fractional* increments in the frequency. Hence musical intervals correspond to frequency *ratios*. For example, an octave corresponds to the frequency ratio of $2:1$, a fifth to a ratio of $3:2$, a fourth to $4:3$, and so on. When we hear a series of notes separated by octaves, their frequencies actually increase in the progression 1, 2, 4, 8, and so on (fig. 36).

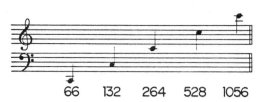

66   132   264   528   1056

FIG. 36. Musical notes separated by equal intervals correspond to frequencies in a geometric progression. The frequencies are in cycles per second.

As a result, the staff on which musical notes are written is actually a logarithmic scale on which vertical distance (pitch) is proportional to the logarithm of the frequency.

The remarkable sensitivity of the human ear to frequency changes is matched by its audibile range—from about 20 cycles per second to about 20,000 (the exact limits vary somewhat with age). In terms of pitch, this corresponds to about ten octaves (an orchestra rarely uses more then seven). By comparison, the eye is sensitive to a wavelength range from 4,000 to 7,000 angstroms (10^{-8}cm)—a range of less than two "octaves."

Among the many phenomena that follow a logarithmic scale, we should also mention the decibel scale of loudness, the brightness scale of stellar magnitudes,[1] and the Richter scale measuring the intensity of earthquakes.

NOTE

1. See, however, John B. Hearnshow, "Origins of the Stellar Magnitude Scale," *Sky and Telescope* (November 1992); Andrew T. Young, "How We Perceive Star Brightnesses," *Sky and Telescope* (March 1990); and S. S. Stevens, "To Honor Fechner and Repeal his Law," *Science* (January 1961).

11

e^{θ}: Spira Mirabilis

Eadem mutata resurgo
(Though changed, I shall arise the same)
—JAKOB BERNOULLI

An air of mystery always surrounds the members of a dynasty. Sibling rivalries, power struggles, and family traits that pass from one generation to the next are the stuff of countless novels and historical romances. England has its royal dynasties, America its Kennedys and Rockefellers. But in the intellectual world it is rare to find a family that, generation after generation, produces creative minds of the highest rank, all in the same field. Two names come to mind: the Bach family in music and the Bernoullis in mathematics.

The ancestors of the Bernoulli family fled Holland in 1583 to escape the Catholic persecution of the Huguenots. They settled in Basel, the quiet university town on the banks of the Rhine where the borders of Switzerland, Germany and France meet. The family members first established themselves as successful merchants, but the younger Bernoullis were irresistibly drawn to science. They were to dominate the mathematical scene in Europe during the closing years of the seventeenth century and throughout most of the eighteenth century.

Inevitably, one compares the Bernoullis with the Bachs. The two families were almost exact contemporaries, and both remained active for some 150 years. But there are also marked differences. In particular, one member of the Bach family stands taller than all the others: Johann Sebastian. His ancestors and his sons were all talented musicians, and some, like Carl Philip Emanuel and Johann Chrisitian, became well-known composers in their own right; but they are all eclipsed by the towering figure of Johann Sebastian Bach.

In the case of the Bernoullis, not one but three figures stand out above the rest: the brothers Jakob and Johann, and the latter's second son, Daniel. Whereas the Bach family lived harmoniously together, with fathers, uncles, and sons all peacefully engaging in the art of music, the Bernoullis were known for their bitter feuds and rival-ries—among themselves as well as with others. By siding with Leib-niz in the priority dispute over the invention of the calculus, they embroiled themselves in numerous controversies. But none of this seems to have had any effect on the vitality of the family; its mem-bers—at least eight achieved mathematical prominence—were blessed with almost inexhaustible creativity, and they contributed to nearly every field of mathematics and physics then known (see fig. 37). And while Johann Sebastian Bach epitomizes the culmination of the Baroque era, bringing to a grand finale a period in music that lasted nearly two centuries, the Bernoullis founded several *new* areas of mathematics, among them the theory of probability and the calcu-lus of variations. Like the Bachs, the Bernoullis were great teachers, and it was through their efforts that the newly invented calculus be-came known throughout continental Europe.

The first of the Bernoullis to achieve mathematical prominence was Jakob (also known as Jacques or James). Born in 1654, he re-ceived a degree in philosophy from the University of Basel in 1671. Rejecting the clerical career his father Nicolaus had intended for him, Jakob pursued his interests in mathematics, physics, and astronomy, declaring, "Against my father's will I study the stars." He traveled and corresponded widely and met some of the leading scientists of the day, among them Robert Hooke and Robert Boyle. From these encounters Jakob learned about the latest developments in physics and astronomy. In 1683 he returned to his native Basel to accept a

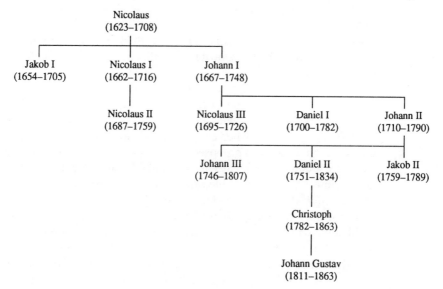

FIG. 37. The Bernoulli family tree.

teaching position at the university, which he held until his death in 1705.

Jakob's second brother, Johann (also known as Johannes, John, or Jeanne) was born in 1667. Like Jakob, he defied his father's wishes to draw him into the family business. He first studied medicine and the humanities, but soon was drawn to mathematics. In 1683 he moved in with Jakob, and from then on their careers were closely tied. Together they studied the newly invented calculus, a task that took them some six years. We must remember that the calculus in those days was an entirely new field, quite difficult to grasp even for professional mathematicians—all the more so because no textbook on the subject had yet been written. So the two brothers had nothing to rely upon except their own perseverence and their active correspondence with Leibniz.

Once they mastered the subject, they undertook to transmit it to others by giving private lessons to several leading mathematicians. Among Johann's students was Guillaume François Antoine de L'Hospital (1661–1704), who then wrote the first calculus textbook, *Analyse des infiniment petits* (Analysis of the infinitely small, published in Paris in 1696). In this work L'Hospital presented a rule to evaluate indeterminate expressions of the form 0/0 (see p. 30). But "L'Hospital's Rule," as it became known (it is now part of the standard calculus course) was actually discovered by Johann. Normally a scientist who publishes under his own name a discovery made by others would be branded as a plagiarist, but in this case it was all done legally, for the two had signed a contract that allowed L'Hosptial, in exchange for the tuition he paid for Johann's tutoring, to use Johann's discoveries as he pleased. L'Hospital's textbook became very popular in Europe and greatly contributed to the spread of the calculus in learned circles.[1]

As the fame of the Bernoulli brothers rose, so did their quarrels. It seems that Jakob became irritated by the success of Johann, while the latter resented the condescending attitude of his older brother. Matters came to a head when each independently solved a problem in mechanics that had been proposed by Johann himself in 1696: to find the curve along which a particle will slide down under the force of gravity in the shortest possible time. This famous problem is known as the *brachistochrone* problem (from the Greek words meaning "shortest time"); already Galileo had tackled it, erroneously believing that the required curve is an arc of a circle. Johann addressed the problem "to the shrewdest mathematicans in all the world" and allowed six months for anyone to come up with a solution. Five correct solutions were submitted—by Newton, Leibniz, L'Hospital, and the two Bernoulli brothers. The required curve turned out to be a *cycloid*,

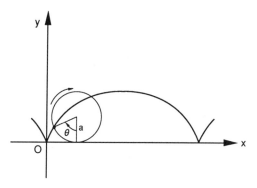

FIG. 38. Cycloid.

the curve traced by a point on the rim of a wheel as it rolls on a horizontal surface (fig. 38).

The graceful shape of this curve and its unique geometric properties had already intrigued several earlier mathematicians. Just a few years before, in 1673, Christian Huygens had found that the cycloid is the solution of another famous problem, that of the *tautochrone*: to find the curve along which a particle moving under the force of gravity will take the same time to reach a given final point, regardless of where the starting point was. (Huygens actually used this result to construct a clock, by constraining the upper end of the pendulum to oscillate between two branches of a cycloid, causing the period to be the same regardless of the amplitude of the oscillations.) Johann was thrilled to discover that the same curve is the solution to both problems: "But you will be petrified with astonishment when I say that exactly this same cycloid, the tautochrone of Huygens, is the brachistochrone we are seeking."[2] But their excitement turned into bitter personal animosity.

Although the two brothers arrived at the same solution independently, they reached it using quite different methods. Johann relied on an analogous problem in optics: to find the curve described by a ray of light as it travels through successive layers of matter of increasing density. The solution makes use of Fermat's Principle, which says that light always follows the path of least time (which is not the same as the path of least distance, a straight line). Today, mathematicians would frown upon a solution that relies heavily on physical principles; but at the end of the seventeenth century the division between pure mathematics and the physical sciences was not taken so seriously, and developments in one discipline strongly influenced the other.

Jakob's approach was more mathematical. He used a new branch of mathematics that he himself had developed: the calculus of variations, an extention of the ordinary calculus. A basic problem in ordinary calculus is to find the values of x that maximize or minimize a given function $y = f(x)$. The calculus of variations extends this problem to finding a *function* that maximizes or minimizes a definite integral (a given area, for example). This problem leads to a certain differential equation whose solution is the required function. The brachistochrone was one of the first problems to which the calculus of variations was applied.

Johann's solution, although correct, used an incorrect derivation. Johann later tried to substitute Jakob's correct derivation as his own. The affair resulted in an exchange of criticism that soon turned ugly. Johann, who held a professorship at the University of Groningen in Holland, vowed not to return to Basel so long as his brother lived. When Jakob died in 1705, Johann accepted his late brother's professorship at the university, which he held until his own death in 1748 at the age of eighty.

To list even superficially the numerous achievements of the Bernoullis would require an entire book.[3] Perhaps Jakob's greatest achievement was his treatise on the theory of probability, the *Ars conjectandi* (The art of conjecture), published posthumously in 1713. This influential work is to the theory of probability what Euclid's *Elements* is to geometry. Jakob also did significant work on infinite series and was the first to deal with the crucial question of convergence. (As we have seen, Newton was aware of this question yet treated infinite series in a purely algebraic manner.) He proved that the series $1/1^2 + 1/2^2 + 1/3^2 + \ldots$ converges but was unable to find its sum (it was not until 1736 that Euler determined it to be $\pi^2/6$). Jakob did important work on differential equations, using them to solve numerous geometric and mechanical problems. He introduced polar coordinates into analytic geometry and used them to describe several spiral-type curves (more about this later). He was the first to use the term *integral calculus* for the branch of the calculus that Leibniz had originally named "the calculus of summation." And Jakob was the first to point out the connection between $\lim_{n\to\infty}(1 + 1/n)^n$ and the problem of continuous compound interest. By expanding the expression $(1 + 1/n)^n$ according to the binomial theorem (see p. 35), he showed that the limit must be between 2 and 3.

Johann Bernoulli's work covered the same general areas as Jakob's: differential equations, mechanics, and astronomy. In the raging Newton-Leibniz controversy, he served as the latter's organ. He also supported the old Cartesian theory of vortices against Newton's more recent gravitational theory. Johann made important contributions to continuum mechanics—elasticity and fluid dynamics—

and in 1738 published his book *Hydraulica*. This work, however, was immediately eclipsed by his son Daniel's treatise *Hydrodynamica*, published in the same year. In this work Daniel (1700–1782) formulated the famous relation between the pressure and velocity of a fluid in motion, a relation known to every student of aerodynamics as Bernoulli's Law; it forms the basis of the theory of flight.

Just as Johann's father Nicolaus had destined a merchant's career for his son, so did Johann himself destine the same career for Daniel. But Daniel was determined to pursue his interests in mathematics and physics. Relations between Johann and Daniel were no better than between Johann and his brother Jakob. Three times Johann won the coveted biennial award of the Paris Academy of Sciences, the third time with his son Daniel (who himself would win it ten times). So embittered was Johann at having to share the prize with his son that he expelled Daniel from his home. Once again the family lived up to its reputation for mixing mathematical excellence with personal feuds.

The Bernoullis continued to be active in mathematics for another hundred years. It was not until the mid-1800s that the family's creativity was finally spent. The last of the mathematical Bernoullis was Johann Gustav (1811–1863), a great-grandson of Daniel's brother Johann II; he died the same year as his father, Christoph (1782–1863). Interestingly, the last of the musical members of the Bach family, Johann Philipp Bach (1752–1846), an organist and painter, also died around that time.

We conclude this brief sketch of the Bernoullis with an anecdote that, like so many stories about great persons, may or may not have happened. While traveling one day, Daniel Bernoulli met a stranger with whom he struck up a lively conversation. After a while he modestly introduced himeslf: "I am Daniel Bernoulli." Upon which the stranger, convinced he was being teased, replied, "and I am Isaac Newton." Daniel was delighted by this unintentional compliment.[4]

Among the many curves that had intrigued mathematicians since Descartes introduced analytic geometry in 1637, two held a special place: the cycloid (mentioned earlier) and the logarithmic spiral. This last was Jakob Bernoulli's favorite; but before we discuss it, we must say a few words about polar coordinates. It was Descartes's idea to locate a point *P* in the plane by giving its distances from two lines (the *x* and *y* axes). But we can also locate *P* by giving its distance *r* from a fixed point *O*, called the *pole* (usually chosen at the origin of the coordinate system) and the angle θ between the line *OP* and a fixed reference line, say the *x*-axis (fig. 39). The two numbers (r, θ)

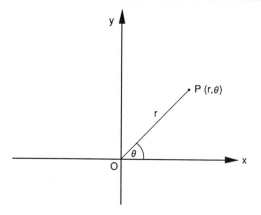

Fig. 39. Polar coordinates.

are the *polar coordinates* of P, just as (x, y) are its rectangular coordinates. At first thought, such a system of coordinates may seem rather strange, but in reality it is quite common—think of how an air traffic controller determines the position of an airplane on the radar screen.

Just as the equation $y = f(x)$ can be interpreted geometrically as the curve described by a moving point with rectangular coordinates (x, y), so can the equation $r = g(\theta)$ be regarded as the curve of a moving point with polar coordinates (r, θ). We should note, however, that the same equation describes quite different curves when interpreted in rectangular or in polar coordinates; for example, the equation $y = 1$ describes a horizontal line, while the equation $r = 1$ describes a circle of radius 1 centered at the origin. And conversely, the same graph has different equations when expressed in rectangular or in polar coordinates: the circle just mentioned has the polar equation $r = 1$ but the rectangular equation $x^2 + y^2 = 1$. Which coordinate system to use is mainly a matter of convenience. Figure 40 shows the 8-shaped curve known as the lemniscate of Bernoulli (named after Jakob), whose polar equation $r^2 = a^2 \cos 2\theta$ is much simpler than the rectangular equation $(x^2 + y^2)^2 = a^2(x^2 - y^2)$.

Polar coordinates were occasionally used before Bernoulli's time,

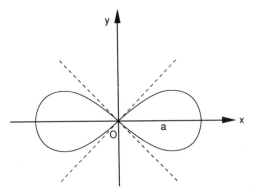

Fig. 40. Lemniscate of Bernoulli.

and Newton, in his *Method of Fluxions*, mentioned them as one of eight different coordinate systems suitable for describing spiral curves. But it was Jakob Bernoulli who first made extensive use of polar coordinates, applying them to a host of curves and finding their various properties. First, however, he had to formulate these properties—the slope of a curve, its curvature, arc length, area, and so on—in terms of polar coordinates, whereas Newton and Leibniz had expressed these properties in terms of rectangular coordinates. Today this is an easy task, given as a routine exercise in a first-year calculus course. In Bernoulli's time it required breaking new ground.

The transformation into polar coordinates enabled Jakob to investigate numerous new curves, which he did with great zest. His favorite curve, as already mentioned, was the logarithmic spiral. Its equation is $\ln r = a\theta$, where a is a constant and ln is the natural or "hyperbolic" logarithm, as it was then called. Today this equation is usually written in reverse, $r = e^{a\theta}$, but in Bernoulli's time the exponential function was not yet regarded as a function in its own right (the number e did not even have a special symbol yet). As is always the practice in calculus, we measure the angle θ not in degrees but in *radians*, that is, in circular measure. One radian is the angle, measured at the center of a circle of radius r, that subtends an arc length equal to r along the circumference (fig. 41). Since the circumference of a circle is $2\pi r$, there are exactly 2π (≈ 6.28) radians in a full rotation; that is, 2π radians = $360°$, from which it follows that one radian is equal to $360°/2\pi$, or approximately $57°$.

If we plot the equation $r = e^{a\theta}$ in polar coordinates, we get the curve shown in figure 42, the logarithmic spiral. The constant a determines the rate of growth of the spiral. If a is positive, the distance r from the pole *increases* as we turn counterclockwise, resulting in a left-handed spiral; if a is negative, r *decreases* and we get a right-handed spiral. The curves $r = e^{a\theta}$ and $r = e^{-a\theta}$ are thus mirror images of each other (fig. 43).

Perhaps the single most important feature of the logarithmic spiral is this: if we increase the angle θ by equal amounts, the distance r from the pole increases by equal *ratios*, that is, in a geometric pro-

[margin annotations:]

$\ln r = a\theta$

or

$r = e^{a\theta}$

One radian

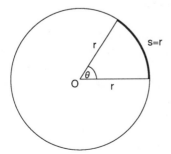

Fig. 41. Radian measure.

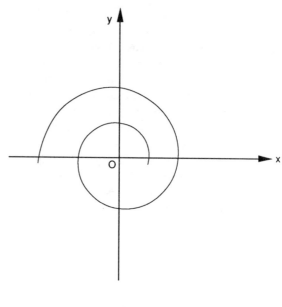

FIG. 42.
Logarithmic
spiral.

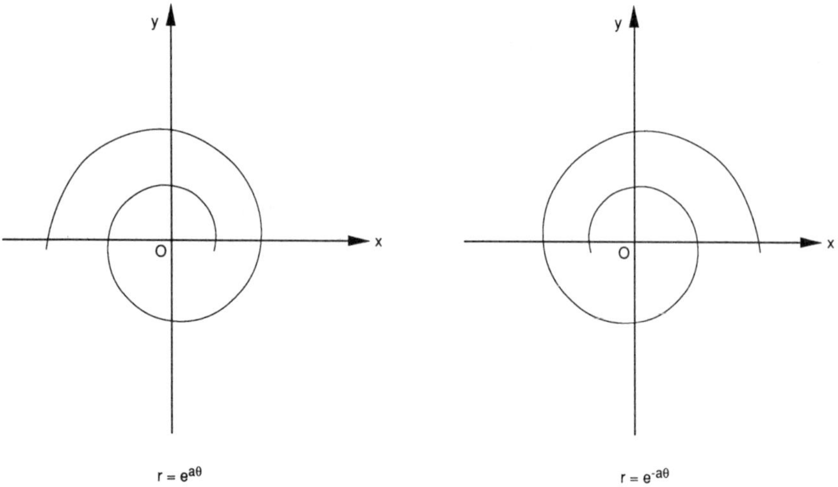

$r = e^{a\theta}$

$r = e^{-a\theta}$

FIG. 43. Left- and right-handed spirals.

gression. This follows from the identity $e^{a(\theta+\varphi)} = e^{a\theta} \cdot e^{a\varphi}$, the factor $e^{a\varphi}$ acting as the common ratio. In particular, if we carry the spiral through a series of full turns (that is, increase θ by multiples of 2π), we can measure the distances along any ray emanating from O and watch their geometric growth.

If we follow the spiral inward from any fixed point P on it, we will have to make an infinite number of rotations before we reach the pole; but surprisingly, the total distance covered is finite. This re-

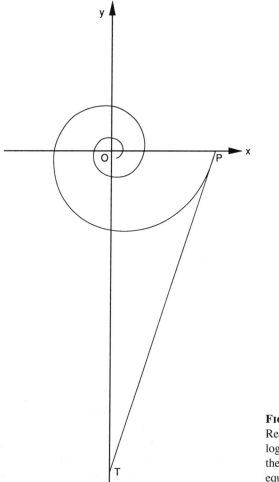

FIG. 44.
Rectification of the
logarithmic sprial:
the distance *PT* is
equal to the arc
length from *P* to *O*.

markable fact was discovered in 1645 by Evangelista Torricelli
(1608–1647), a disciple of Galileo who is known mainly for his experiments in physics. He showed that the arc length from *P* to the
pole is equal to the length of the tangent line to the spiral at *P*, measured between *P* and the *y*-axis (fig. 44). Torricelli treated the spiral
as a succession of radii increasing in a geometric progression as θ
increases arithmetically, reminiscent of Fermat's technique in finding
the area under the curve $y = x^n$. (With the help of the integral calculus, of course, the result is much simpler to obtain; see Appendix 6.)
His result was the first known *rectification*—finding the length of
arc—of a non-algebraic curve.

Some of the most remarkable properties of the logarithmic spiral
depend on the fact that the function e^x is equal to its own derivative.

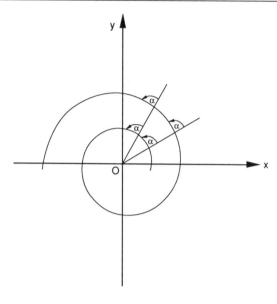

FIG. 45.
Equiangular
property of the
logarithmic sprial:
every line through
the pole O
intersects the spiral
at the same angle.

For example, *every straight line through the pole intersects the spiral at the same angle* (fig. 45; a proof of this property is given in Appendix 6). Moreover, the logarithmic spiral is the *only* curve that has this property; hence it is also known as the *equiangular spiral*. This makes the spiral a close relative of the circle, for which the angle of intersection is 90°. Indeed, the circle is a logarithmic spiral whose rate of growth is 0: putting $a = 0$ in the equation $r = e^{a\theta}$, we get $r = e^0 = 1$, the polar equation of the unit circle.

What excited Jakob Bernoulli most about the logarithmic spiral is the fact that it remains *invariant*—unchanged—under most of the transformations of geometry. Consider, for example, the transformation of inversion. A point P whose polar coordinates are (r, θ) is "mapped" onto a point Q with polar coordinates $(1/r, \theta)$ (fig. 46). Usually, the shape of a curve changes drastically under inversion; for

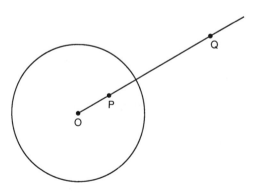

FIG. 46. Inversion in
the unit circle:
$OP \cdot OQ = 1$.

example, the hyperbola $y = 1/x$ is transformed into the lemniscate of Bernoulli mentioned earlier. This is not surprising, since changing r into $1/r$ means that points very close to O go over to points very far from it, and vice versa. But not so with the logarithmic spiral: changing r into $1/r$ merely changes the equation $r = e^{a\theta}$ into $r = 1/e^{a\theta} = e^{-a\theta}$, whose graph is a mirror image of the original spiral.

evolute

Just as inversion transforms a given curve into a new one, so we can obtain a new curve by constructing the *evolute* of the original curve. This concept involves the center of curvature of the curve. As mentioned earlier, the *curvature* at each point of a curve is a measure of the rate at which the curve changes direction at that point; it is a number that varies from point to point (just as the slope of a curve changes from point to point) and is therefore a function of the independent variable. The curvature is denoted by the Greek letter κ (kappa); its reciprocal, $1/\kappa$, is called the *radius of curvature* and is denoted by the letter ρ (rho). The smaller ρ is, the greater the curvature at that point, and vice versa. A straight line has a curvature of 0, hence its radius of curvature is infinite. A circle has a constant curvature, and its radius of curvature is simply its radius.

If we draw a perpendicular to the tangent line at each point of a curve (on the concave side) and along it measure a distance equal to the radius of curvature at that point, we arrive at the *center of curvature* of that point (fig. 47). The evolute is the locus of the centers of curvature of the original curve as we move along it. Usually, the evolute is a new curve, different from the one from which it was generated; for example, the evolute of the parabola $y = x^2$ is a semicubical parabola, a curve whose equation is of the form $y = x^{2/3}$ (fig. 48). But as Jakob Bernoulli found to his great delight, the logarithmic spiral is its own evolute. (The cycloid, too, has this property; but the evolute

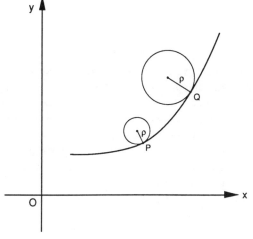

FIG. 47. Radius and center of curvature.

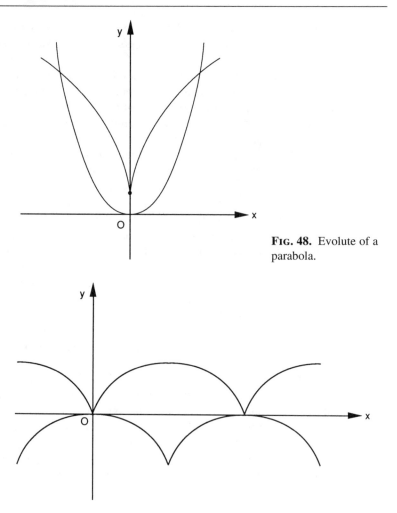

FIG. 48. Evolute of a parabola.

FIG. 49. The evolute of a cycloid is an idential cycloid but shifted relative to the first.

of a cycloid is a second cycloid, identical to the first but shifted with respect to it [fig. 49], whereas the evolute of a logarithmic spiral is the same spiral.) He also discovered that the *pedal curve* of a logarithmic spiral—the locus of the perpendicular projections from the pole to the tangent lines of the given curve—is again the same spiral. And if that was not enough, he found that the *caustic* of a logarithmic spiral—the envelope formed by rays of light emanating from the pole and reflected by the curve—is again the same spiral.

So moved was Jakob by these discoveries that he developed an almost mystical reverence toward his beloved curve: "Since this marvellous spiral, by such a singular and wonderful peculiarity . . . always produces a spiral similar to itself, indeed precisely the same spiral, however it may be involved or evolved, or reflected or refracted . . . it may be used as a symbol, either of fortitude and constancy in adversity, or of the human body, which after all its changes, even after death, will be restored to its exact and perfect self."[5] He dubbed it *spira mirabilis* (the marvelous spiral) and expressed his wish that a logarithmic spiral be engraved on his tombstone with the inscription, *Eadem mutata resurgo* (Though changed, I shall arise the same), in the tradition of Archimedes, who, according to legend, asked that a sphere with a circumscribed cylinder be engraved on his tomb. Jakob's wish was fulfilled—almost. Whether out of ignorance or to make his task easier, the mason indeed cut a spiral on the grave, but it was an Archimedean instead of a logarithmic spiral. (In an Archimedean, or linear, spiral each successive turn increases the distance from the pole by a constant *difference* rather than ratio; the sound grooves on a vinyl record follow a linear spiral.) Visitors to the cloisters at the Münster cathedral in Basel can still see the result (fig. 50), which no doubt would have made Jakob turn in his grave.

NOTES AND SOURCES

1. See Chapter 9, note 9.

2. Quoted in Eric Temple Bell, *Men of Mathematics*, 2 vols. (1937; rpt. Harmondsworth: Penguin Books, 1965), 1:146.

3. The Swiss publishing house Birkhäuser has undertaken the publication of the Bernoulli family's scientific work and correspondence. This monumental task, begun in 1980 and scheduled for completion in 2000, will encompass at least thirty volumes.

4. Bell, *Men of Mathematics*, 1:150; also Robert Edouard Moritz, *On Mathematics and Mathematicians (Memorabilia Mathematica)* (1914; rpt. New York: Dover, 1942), p. 143.

5. Quoted in Thomas Hill, *The Uses of Mathesis*, Bibliotheca Sacra, vol. 32, pp. 515–516, as quoted by Moritz, *On Mathematics and Mathematicians*, pp. 144–145.

Birkhäuser
is Swiss

FIG. 50. Jakob Bernoulli's tombstone in Basel. Reproduced with permission from Birkhäuser Verlag AG, Basel.

A Historic Meeting between J. S. Bach and Johann Bernoulli

Did any member of the Bach family ever meet one of the Bernoullis? It's unlikely. Travel in the seventeenth century was an enterprise to be undertaken only for compelling reasons. Barring a chance encounter, the only imaginable reason for such a meeting would have been an intense curiosity about the activities of the other, and there is no evidence of that. Nevertheless, the thought that perhaps such an encounter did take place is compelling. Let us imagine a meeting between Johann Bernoulli (Johann I, that is) and Johann Sebastian Bach. The year is 1740. Each is at the peak of his fame. Bach, at the age of fifty-five, is organist, composer, and Kapellmeister (musical director) at St. Thomas's Church in Leipzig. Bernoulli, at seventy-three, is the most distinguished professor of the University of Basel. The meeting takes place in Nuremberg, about halfway between their home towns.

BACH: Herr Professor, I am very glad to meet you at last, having heard so much about your remarkable achievements.

BERNOULLI: I am equally delighted to meet you, Herr Kapellmeister. Your fame as an organist and composer has reached far beyond the Rhine. But tell me, are you really interested in my work? I mean, musicians are not usually versed in mathematics, are they? And truth be told, my interest in music is entirely theoretical; for example, a while ago I and my son Daniel did some studies on the theory of the vibrating string. This is a new field of research involving what we in mathematics call continuum mechanics.[1]

BACH: In fact, I too have been interested in the way a string vibrates. As you know, I also play the harpsichord, whose sound is produced by plucking the strings through the action of the keys. For years I have been bothered by a technical problem with this instrument, which I have been able to solve only recently.

BERNOULLI: And what is that?

BACH: As you know, our common musical scale is based on the laws of the vibrating string. The intervals we use in music—the octave, fifth, fourth, and so on—are all derived from the harmonics, or overtones, of a string—those feeble higher tones that are always present when a string vibrates. The frequencies of these harmonics are integral multiples of the fundamental (lowest) frequency, so they

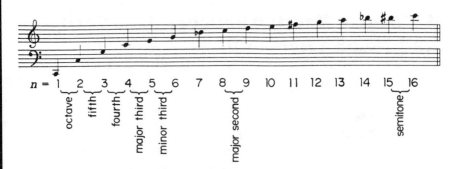

FIG. 51. The series of harmonics, or overtones, emitted by a vibrating string. The numbers indicate the relative frequencies of the notes.

form the progression 1, 2, 3, 4, . . . [fig. 51]. The intervals of our scale correspond to *ratios* of these numbers: 2:1 for the octave, 3:2 for the fifth, 4:3 for the fourth, and so on. The scale formed from these ratios is called the *just intonation scale*.

BERNOULLI: That perfectly fits my love for orderly sequences of numbers.

BACH: But there is a problem. A scale constructed from these ratios consists of three basic intervals: 9:8, 10:9, and 16:15 [fig. 52]. The first two are nearly identical, and each is called a whole tone, or a *second* (so named because it leads to the second note in the scale). The last ratio is much smaller and is called a *semitone*. Now, if you start with the note C and go up the scale C-D-E-F-G-A-B-C′, the first interval, from C to D, is a whole tone whose frequency ratio is 9:8. The next interval, from D to E, is again a whole tone, but its frequency ratio is 10:9. The remaining intervals in the scale are E to F (16:15), F to G (9:8), G to A (10:9), A to B (9:8), and finaly B to C′ (16:15)—the last note being one octave above C. This is the scale known as C-major. But the same ratios should hold regardless of which note we start from. *Every* major scale consists of the same sequence of intervals.

BERNOULLI: I can see the confusion of having two different ra-

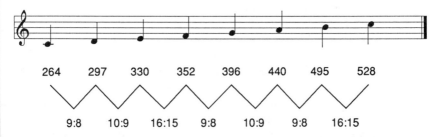

FIG. 52. The scale of C-major. The upper numbers indicate the frequency of each note in cycles per second; the lower numbers are the frequency ratios between successive notes.

tios for the same interval. But why does this trouble you? After all, music has been around for many centuries, and no one else has been bothered.

BACH: Actually, it's worse than that. Not only are there two different kinds of whole tones in use, but if we add up two semitones, their sum will not exactly equal either of the whole tones. You can figure it out yourself. It's as if 1/2 + 1/2 were not exactly equal to 1, only approximately.

BERNOULLI (*jotting down some figures in his notebook*): You're right. To add two intervals, we must multiply their frequency ratios. Adding two semitones corresponds to the product $(16:15) \cdot (16:15) = 256:225$ or approximately 1.138, which is slightly greater than either $9:8$ (= 1.125) or $10:9$ (= 1.111).

BACH: You see what happens. The harpsichord has a delicate mechanism that allows each string to vibrate only at a specific fundamental frequency. This means that if I want to play a piece in D-major instead of C-major—what is known as transposition—then the first interval (from D to E) will have the ratio $10:9$ instead of the original $9:8$. This is still all right, because the ratio $10:9$ is still part of the scale; and besides, the average listener can barely tell the difference. But the next interval—which must again be a whole tone—can be formed only by going up a semitone from E to F and then another semitone from F to F-sharp. This corresponds to a ratio of $(16:15) \cdot (16:15) = 256:225$, an interval that does not exist in the scale. And the problem is only compounded the farther up I go in the new scale. In short, with the present system of tuning I cannot transpose from one scale to another, unless of course I happen to play one of those few instruments that have a continuous range of notes, such as the violin or the human voice.

BACH (*not waiting for Bernoulli to respond*): But I have found a remedy: I make all whole tones equal to one another. This means that any two semitones will always add up to a whole tone. But to accomplish this I had to abandon the just intonation scale in favor of a compromise. In the new arrangement, the octave consists of twelve *equal semitones*. I call it the *equal-tempered scale*.[2] The problem is, I have a hard time convincing my fellow musicians of its advantages. They cling stubbornly to the old scale.

BERNOULLI: Perhaps I can help you. First of all, I need to know the frequencey ratio of each semitone in your new scale.

BACH: Well, you're the mathematician; I'm sure you can figure it out.

BERNOULLI: I just did. If there are twelve equal semitones in the octave, then each semitone must have a frequency ratio of $\sqrt[12]{2} : 1$. Indeed, adding twelve of these semitones corresponds to $(\sqrt[12]{2})^{12}$, which is exactly $2:1$, the octave.[3]

BACH: Now you've completely lost me. My knowledge of math barely goes beyond elementary arithmetic. Is there any way you could demonstrate this visually?

BERNOULLI: I think I can. My late brother Jakob spent much time exploring a curve called the logarithmic spiral. In this curve, equal rotations increase the distance from the pole by equal *ratios*. Isn't this exactly the case in the scale you've just described to me?

BACH: Can you show me this curve?

BERNOULLI: Sure [fig. 53]. While you were talking, I marked on it the twelve equal semitones. To transpose a piece from one scale to another, all you have to do is turn the spiral so that the first tone of your scale falls on the *x*-axis. The remaining tones will automatically fall into place. It's really a musical calculator of sorts!

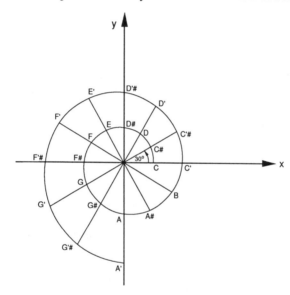

FIG. 53.
The twelve notes of the equal-tempered scale arranged along a logarithmic sprial.

BACH: This sounds exciting. Perhaps your spiral can help me teach the subject to younger musicians, because I am convinced that the new scale holds great promise for future performers. Indeed, I am now working on a series of preludes that I call "The Well-Tempered Clavier." Each prelude is written in one of the twelve major and twelve minor keys. I wrote a similar series in 1722 and intended it as an instruction book for my first wife, Maria Barbara—may she rest in peace—and my first son, Wilhelm Friedemann. Since then, as you know, I have been blessed with many more children, all of whom show signs of great musical talent. It is for them, as well as for my second wife, Anna Magdalena, that I am writing this new work.

BERNOULLI: I admire the wonderful relations you have with your children. Unfortunately, I cannot say the same about my own family. For some reason we have always been a quarrelsome lot. I mentioned to you my son Daniel, with whom I worked on several problems. But six years ago I had to share with him the biennial award of the Paris Academy of Sciences. I felt that the prize really should have been mine alone. Furthermore, Daniel has always been on Newton's side in his bitter controversy with Leibniz, while I have steadfastly supported Leibniz, whom I regard as the true inventor of the calculus. Under these circumstances, I found it impossible to continue my work with him, and I have ordered him out of my house.

BACH (*hardly able to hide his amazement*): Well, I wish you and your family my very best, and let God bless you with many more years of productive life.

BERNOULLI: I wish you the same. And God willing, may we meet again and continue our dialogue, now that we have discovered that mathematics and music have so much in common.

The two shake hands and depart for their long journeys home.

NOTES

1. The vibrating string was the outstanding problem in mathematical physics throughout the eighteenth century. Most of the leading mathematicians of the period contributed to its solution, among them the Bernoullis, Euler, D'Alembert, and Lagrange. The problem was finally solved in 1822 by Joseph Fourier.

2. Bach was not the first to think of such an arrangement of notes. Attempts to arrive at a system of "correct" tuning had been made as early as the sixteenth century, and in 1691 a "well-tempered" scale was suggested by the organ builder Andreas Werckmeister. It was owing to Bach, however, that the equal-tempered scale became universally known. See *The New Grove Dictionary of Music and Musicians*, vol. 18 (London: Macmillan, 1980), pp. 664–666 and 669–670.

3. The decimal value of this ratio is about 1.059, compared to 1.067 for the ratio 16:15. This slight difference, though still within the range of audibility, is so small that most listeners ignore it. When playing solo, however, singers and string instrumentalists still prefer to use the just intonation scale.

The Logarithmic Spiral in
Art and Nature

Probably no curve has had greater appeal for scientists, artists, and naturalists than the logarithmic spiral. Dubbed *spira mirabilis* by Jakob Bernoulli, the spiral possesses remarkable mathematical properties that make it unique among plane curves (see p. 121). Its graceful shape has been a favorite decorative motif since antiquity; and, with the possible exception of the circle (which itself is a special case of a logarithmic spiral), it occurs more often in nature than any other curve, sometimes with stunning accuracy, as in the nautilus shell (fig. 54).

Perhaps the most remarkable fact about the logarithmic spiral is that it looks the same in all directions. More precisely, every straight line through the center (pole) intersects the spiral at exactly the same angle (see fig. 45 in Chapter 11). Hence it is also known as the *equiangular* spiral. This property endows the spiral with the perfect symmetry of the circle—indeed, the circle is a logarithmic spiral for which the angle of intersection is 90° and the rate of growth is 0.

A second feature, related to the first, is this: rotating the spiral by equal amounts increases the distance from the pole by equal *ratios*, that is, in a geometric progression. Hence, any pair of lines through the pole with a fixed angle between them cut similar (though not congruent) sectors from the spiral. This is clearly seen in the nautilus

Fig. 54.
Nautilus shell.

shell, whose chambers are precise replicas of one another, increasing geometrically in size. In his classic work *On Growth and Form*, the English naturalist D'Arcy W. Thompson (1860–1948) discusses in great detail the role of the logarithmic spiral as the preferred growth pattern of numerous natural forms, among them shells, horns, tusks, and sunflowers (fig. 55).[1] To these we may add spiral galaxies, those "island universes" whose precise nature was not yet known when Thompson published his book in 1917 (fig. 56).

The early years of the twentieth century saw a revival of interest in Greek art and its relation to mathematics; theories of aesthetics abounded, and some scholars attempted to give the concept of beauty a mathematical formulation. This led to a rediscovery of the logarithmic spiral. In 1914 Sir Theodore Andrea Cook published *The Curves of Life*, a work of nearly five hundred pages devoted entirely to the spiral and its role in art and nature. Jay Hambidge's *Dynamic Symme-*

FIG. 55. Sunflower.

FIG. 56. The spiral galaxy M100. Courtesy of Zsolt Frei.

try (1926) influenced generations of artists striving for perfect beauty and harmony. Hambidge used as his guiding principle the *golden ratio*, the ratio in which a line segment must be divided so that the entire length is to the long part as the long part is to the short (fig. 57). This ratio, denoted by the Greek letter ϕ (phi), has the value $(1 + \sqrt{5})/2 = 1.618 \ldots$. Many artists believe that of all rectangles, the one with a length-to-width ratio equal to ϕ—the "golden rectangle"—has the "most pleasing" dimensions; hence the prominent role this ratio has played in architecture. From any golden rectangle one can get a new golden rectangle whose length is the width of the original rectangle. This process can be repeated indefinitely, resulting in an infinite sequence of golden rectangles whose sizes shrink to 0 (fig. 58). These rectangles circumscribe a logarithmic spiral, the "golden spiral," which Hambidge used as his motif. One author influenced by Hambidge's ideas was Edward B. Edwards, whose *Pattern and Design with Dynamic Symmetry* (1932) presents hundreds of decorative designs based on the spiral motif (fig. 59).

The Dutch artist Maurits C. Escher (1898–1972) used the spiral in some of his most creative works. In *Path of Life* (1958; fig. 60) we see a grid of logarithmic spirals along which fish swim in an endless cycle. Emerging from the infinitely remote center, they are white; but

FIG. 57. The golden ratio: *C* divides the segment *AB* such that the whole segment is to the large part as the large part is to the small. If the whole segment is of unit length, we have $1/x = x/(1 - x)$. This leads to the quadratic equation $x^2 + x - 1 = 0$, whose positive solution is $x = (-1 + \sqrt{5})/2$, or about 0.61803. The golden ratio is the reciprocal of this number, or about 1.61803.

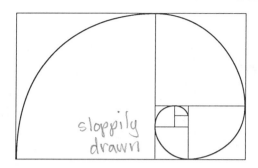

Here we have an infinite series definitely equal

sloppily drawn

FIG. 58. "Golden rectangles" inscribed in a logarithmic spiral. Each rectangle has a length-to-width ratio of 1.61803. . . .

to a finite area by geometric construction. The succession of diminishing squares completely covers the containing rectangle.

FIG. 59. Decorative patterns based on the logarithmic spiral. Reprinted from Edward B. Edwards, *Pattern and Design with Dynamic Symmetry* (1932; New York: Dover, 1967), with permission.

FIG. 60. M.C. Escher, *Path of Life II* (1958). Copyright © M.C. Escher / Cordon Art -Baarn - Holland. All rights reserved.

as they near the periphery, their color changes to gray, whence they move back to the center and disappear there—the eternal cycle of life and death. Escher's passion for filling a plane with figures of identical shape whose sizes increase geometrically finds here a sublime expression.[2]

Imagine four bugs positioned at the corners of a rectangle. At the sound of a signal, each bug starts to move toward its neighbor. What paths will they follow, and where will they meet? The paths turn out to be logarithmic spirals that converge at the center. Figure 61 shows one of many designs based on the Four Bug Problem.

Here's a thought for those who like to dream about "what would happen if . . . " Had the universal law of gravitation been an inverse *cubic* law instead of an inverse square law, one possible orbit of the planets around the sun would be a logarithmic spiral (the hyperbolic spiral $r = k/\theta$ would be another possible orbit). This was proved by Isaac Newton in Book I of his *Prinicipia*.

FIG. 61. Decorative design based on the Four Bug Problem.

NOTES AND SOURCES

1. All the works cited in this chapter are listed in the Bibliography.

2. For a detailed discussion of the logarithmic spiral in Escher's work, see my book *To Infinity and Beyond: A Cultural History of the Infinite* (1987; rpt. Princeton: Princeton University Press, 1991).

12

$(e^x + e^{-x})/2$: The Hanging Chain

Therefore, I have attacked [the problem of the catenary],

which I had hitherto not attempted, and with my key

[the differential calculus] happily opened its secret.

—Gottfried Wilhelm Leibniz, in *Acta eruditorum*

(July 1690)

We are not quite done with the Bernoullis yet. Among the outstanding problems that occupied the mathematical community in the decades following the invention of the calculus was the problem of the *catenary*—the hanging chain (from the Latin *catena*, a chain). This problem, like the brachistochrone, was first proposed by one of the Bernoulli brothers, this time Jakob. In the May 1690 issue of *Acta eruditorum*, the journal that Leibniz had founded eight years earlier, Jakob wrote: "And now let this problem be proposed: To find the curve assumed by a loose string hung freely from two fixed points."[1] Jakob assumed that the string is flexible in all its parts and that it has a constant thickness (and thus a uniform linear density).

The history of this celebrated problem closely parallels that of the brachistochrone, and most of the same players took part. Galileo had already shown interest in it and thought that the required curve is a parabola. To the eye, a hanging chain certainly looks like a parabola (fig. 62). But Christian Huygens (1629–1695), the prolific Dutch scientist whose place in history has always been somewhat underrated (no doubt because he lived between the eras of Kepler and Galileo before him and Newton and Leibniz after him), proved that the catenary could not possibly be a parabola. This was in 1646, when Huygens was only seventeen years old. But to find the actual curve was another matter, and at the time no one had any idea how to tackle the problem. It was one of nature's great mysteries, and only the calculus could possibly solve it.

In June 1691, one year after Jakob Bernoulli proposed his problem, the *Acta* published the three correct solutions that were submitted—by Huygens (now sixty-two years old), Leibniz, and Johann Bernoulli. Each attacked the problem differently, but all arrived at the

FIG. 62. The catenary: the curve of a hanging chain.

same solution. Jakob himself was unable to solve it, which delighted his brother Johann all the more. Twenty-seven years later, long after Jakob's death, Johann wrote to a colleague who had apparently questioned Johann's claim that he, and not Jakob, had found the solution:

You say that my brother proposed this problem; that is true, but does it follow that he had a solution of it then? Not at all. When he proposed this problem at my suggestion (for I was the first to think of it), neither the one nor the other of us was able to solve it; we despaired of it as insoluble, until Mr. Leibniz gave notice to the public in the Leipzig journal of 1690, p. 360, that he had solved the problem but did not publish his solution, so as to give time to other analysts, and it was this that encouraged us, my brother and me, to apply ourselves afresh.

The efforts of my brother were without success; for my part, I was more fortunate, for I found the skill (I say it without boasting, why should I conceal the truth?) to solve it in full. . . . The next morning, filled with Joy, I ran to my brother, who was still struggling miserably with this Gordian knot without getting anywhere, always thinking like Galileo that the catenary was a parabola. Stop! Stop! I say to him, don't torture yourself anymore to try to prove the identity of the catenary with the parabola, since it is entirely false.[2]

Johann added that, of the two curves, the parabola is algebraic, while the catenary is transcendental. Boisterous as always, Johann concluded: "You knew the disposition of my brother. He would sooner have taken away from me, if he could have done so honestly, the honor of being the first to solve it, rather than letting me take part by myself, let alone ceding me the place, if it had really been his." The Bernoullis' notoriety for feuding amongst themselves—and with others—did not diminish a bit with the passage of time.[3]

The catenary turned out to be a curve whose equation, in modern notation, is $y = (e^{ax} + e^{-ax})/2a$, where a is a constant whose value depends on the physical parameters of the chain—its linear density (mass per unit length) and the tension at which it is held. The discovery of this equation was hailed as a great triumph of the new differential calculus, and the contestants made the most of it to advance their reputations. For Johann, it was the "passport to enter the learned society of Paris."[4] Leibniz saw to it that everyone knew it was *his* calculus (his "key") that solved the mystery. If such boasting sounds excessive today, we should remember that in the closing years of the seventeenth century problems like the brachistochrone and the catenary presented the utmost challenge to mathematicians, and their solutions were justly regarded with great pride. Today these problems are routine exercises in an advanced calculus course.[5]

We should mention that the equation of the catenary was not originally given in the above form. The number e did not have a special symbol yet, and the exponential function was regarded not as a function in its own right but as the inverse of the logarithmic function. The equation of the catenary was simply implied from the way it was constructed, as Leibniz's own drawing (fig. 63) clearly shows. Leibniz even suggested that the catenary could be used as a device for calculating logarithms, an "analog" logarithmic table of sorts. "This may help," he said, "since on long trips one may lose his table of logarithms."[6] Was he suggesting that one should carry a chain in his pocket as a backup logarithmic table?

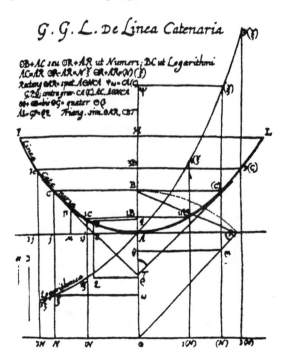

FIG. 63. Leibniz's construction of the catenary (1690).

FIG. 64. The Gateway Arch, St. Louis, Missouri. Courtesy of the Jefferson National Expansion Memorial / National Park Service.

In our century the catenary has been immortalized in one of the world's most imposing architectural monuments, the Gateway Arch in St. Louis, Missouri (fig. 64). Designed by the architect Eero Saarinen and completed in 1965, it has the precise shape of an inverted catenary, its top towering 630 feet above the banks of the Mississippi River.

✧   ✧   ✧

For $a = 1$ the equation of the catenary is

$$y = \frac{e^x + e^{-x}}{2}. \tag{1}$$

Its graph can be constructed by plotting the graphs of e^x and e^{-x} on the same coordinate system, adding their ordinates (heights) for every

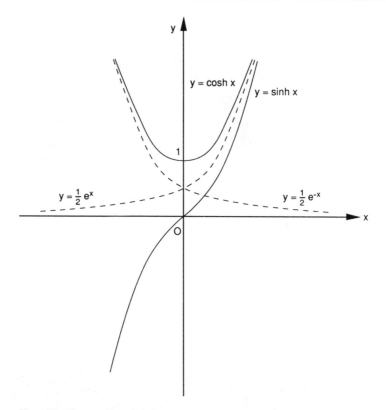

FIG. 65. The graphs of sinh x and cosh x.

point x, and dividing the result by 2. The graph, which by its very manner of construction is symmetric about the y-axis, is shown in figure 65.

In addition to equation 1 we may consider a second equation,

$$y = \frac{e^x - e^{-x}}{2},$$ (2)

whose graph is also shown in figure 65. It so happens that equations 1 and 2, when regarded as functions of x, exhibit some striking similarities to the circular functions $\cos x$ and $\sin x$ studied in trigonometry. These similarities were first noticed by the Italian Jesuit Vincenzo Riccati (1707–1775). In 1757 he introduced the notation Ch x and Sh x for these functions:

$$\text{Ch } x = \frac{e^x + e^{-x}}{2}, \quad \text{Sh } x = \frac{e^x - e^{-x}}{2}.$$ (3)

He showed that they satisfy the identity (Ch φ)^2 − (Sh φ)^2 = 1 (where we have used the letter φ for the independent variable), which, except for the minus sign of the second term, is analogous to the trigonometric identity $(\cos \varphi)^2 + (\sin \varphi)^2 = 1$. This shows that Ch φ and Sh φ are related to the hyperbola $x^2 - y^2 = 1$ in the same way as $\cos \varphi$ and $\sin \varphi$ are related to the unit circle $x^2 + y^2 = 1$.[7] Riccati's notation has survived almost unchanged; today we denote these functions by cosh φ and sinh φ—read "hyperbolic cosine of φ" and "hyperbolic sine of φ" (the former is sometimes pronounced the way it is written, "cosh" [as in "posh"], but this is a bit awkward with sinh).

Riccati belonged to yet another remarkable family of mathematicians, though one not as prolific as the Bernoullis. Vincenzo's father, Jacopo (or Giacomo) Riccati (1676–1754), had studied at the University of Padua and later did much to disseminate Newton's work in Italy (the differential equation $dy/dx = py^2 + qy + r$, where p, q, and r are given functions of x, is named after Jacopo Riccati). Two other of Jacopo's sons, Giordano (1709–1790) and Francesco (1718–1791), also became successful mathematicians, the latter applying geometric principles to architecture. Vincenzo Riccati was intrigued by the similarity between the equations $x^2 - y^2 = 1$ and $x^2 + y^2 = 1$ of the hyperbola and the unit circle. He developed his theory of hyperbolic functions entirely from the geometry of the hyperbola. Today we prefer the analytic approach, which makes use of the special properties of the functions e^x and e^{-x}. For example, the identity $(\cosh \varphi)^2 - (\sinh \varphi)^2 = 1$ can easily be proved by squaring the right sides of equations 3, subtracting the result, and using the identities $e^x \cdot e^y = e^{x+y}$ and $e^0 = 1$.

Good

It turns out that most of the formulas of ordinary trigonometry have their hyperbolic counterparts. That is, if we take a typical trigonometric identity and replace sin φ and cos φ by sinh φ and cosh φ, the identity will still be correct, with a possible change of sign in one or more terms. For example, the circular functions obey the differentiation formulas

$$\frac{d}{dx}(\cos x) = -\sin x, \quad \frac{d}{dx}(\sin x) = \cos x. \tag{4}$$

The corresponding formulas for the hyperbolic functions are

$$\frac{d}{dx}(\cosh x) = \sinh x, \quad \frac{d}{dx}(\sinh x) = \cosh x \tag{5}$$

(note the absence of the minus sign in the first of equations 5). These similarities make the hyperbolic functions useful in evaluating certain indefinite integrals (antiderivatives), for example, integrals of the form $(a^2 + x^2)^{1/2}$. (A list of some additional analogies between the circular and hyperbolic functions can be found on page 148).

One would wish that *every* relation among the circular functions had its hyperbolic counterpart. This would put the circular and hyperbolic functions on a completely equal basis, and by implication give the hyperbola a status equal to that of the circle. Unfortunately, this is not the case. Unlike the hyperbola, the circle is a closed curve; as we go around it, things must return to their original state. Consequently, the circular functions are *periodic*—their values repeat every 2π radians. It is this feature that makes the circular functions central to the study of periodic phenomena—from the analysis of musical sounds to the propagation of electromagnetic waves. The hyperbolic functions lack this feature, and their role in mathematics is less fundamental.[8]

Yet in mathematics, purely formal relations often have great suggestive power and have motivated the development of new concepts. In the next two chapters we shall see how Leonhard Euler, by allowing the variable x in the exponential function to assume imaginary values, put the relations between the circular and the hyperbolic functions on an entirely new foundation.

NOTES AND SOURCES

1. Quoted in C. Truesdell, *The Rational Mechanics of Flexible or Elastic Bodies, 1638–1788* (Switzerland: Orell Füssli Turici, 1960), p. 64. This work also contains the three derivations of the catenary as given by Huygens, Leibniz, and Johann Bernoulli.

2. Ibid., pp. 75–76.

3. For the sake of fairness, we should mention that Jakob extended Johann's method of solution to chains with variable thickness. He also proved that of all possible shapes a hanging chain can assume, the catenary is the one with the lowest center of gravity—an indication that nature strives to minimize the potential energy of the shapes it creates.

4. Ludwig Otto Spiess, as quoted in Truesdell, *Rational Mechanics*, p. 66.

5. For the solution of the catenary problem, see, for example, George F. Simmons, *Calculus with Analytic Geometry* (New York: McGraw-Hill, 1985), pp. 716–717.

6. Quoted in Truesdell, *Rational Mechanics*, p. 69.

7. Note, however, that for the hyperbolic functions the variable φ no longer plays the role of an angle, as is the case with the circular functions. For a geometric interpretation of φ in this case, see Appendix 7.

8. In Chapter 14, however, we will see that the hyperbolic funtions have an *imaginary* period $2\pi i$, where $i = \sqrt{-1}$.

Remarkable Analogies

Consider the unit circle—the circle with center at the origin and radius 1—whose equation in rectangular coordinates is $x^2 + y^2 = 1$ (fig. 66). Let $P(x, y)$ be a point on this circle, and let the angle between the positive x-axis and the line OP be φ (measured counterclockwise in radians). The *circular* or *trigonometric functions* "sine" and "cosine" are defined as the x and y coordinates of P:

$$x = \cos \varphi, \quad y = \sin \varphi.$$

I don't get this at all.

The angle φ can also be interpreted as twice the area of the circular sector OPR in figure 66, since this area is given by the formula $A = r^2\varphi/2 = \varphi/2$, where $r = 1$ is the radius.

The *hyperbolic functions* are similarly defined in relation to the rectangular hyperbola $x^2 - y^2 = 1$ (fig. 67), whose graph can be obtained from the hyperbola $2xy = 1$ by rotating the coordinate axes

Area of ⊙ = πr^2

circumference of ⊙ = $2\pi r$

When $r = 1$, circumference = 2π

So $A = r^2 \dfrac{\varphi}{2}$

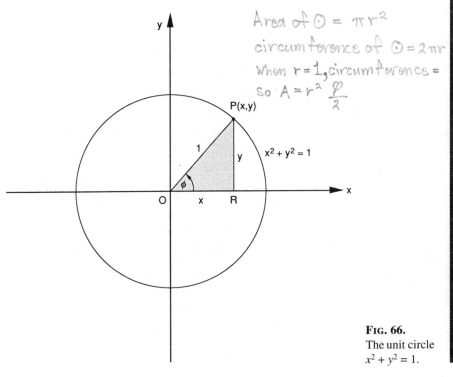

$P(x,y)$

$x^2 + y^2 = 1$

FIG. 66.
The unit circle
$x^2 + y^2 = 1$.

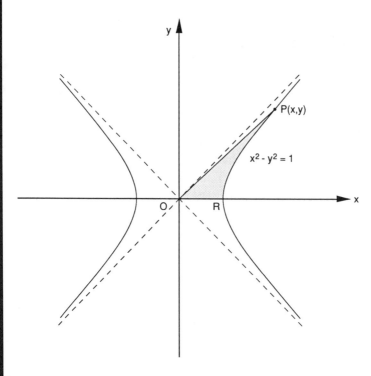

through 45° counterclockwise; it has the pair of lines $y = \pm x$ as asymptotes. Let $P(x, y)$ be a point on this hyperbola. We then define:

$$x = \cosh \varphi, \quad y = \sinh \varphi,$$

where $\cosh \varphi = (e^\varphi + e^{-\varphi})/2$ and $\sinh \varphi = (e^\varphi - e^{-\varphi})/2$ (see p. 144). Here φ is *not* the angle between the x-axis and the line OP, but merely a parameter (variable).

Below, listed side by side, are several analogous properties of the circular and hyperbolic functions (we use x for the independent variable):

Pythagorean Relations

$$\cos^2 x + \sin^2 x = 1 \qquad \cosh^2 x - \sinh^2 x = 1$$

Here $\cos^2 x$ is short for $(\cos x)^2$, and similarly for the other functions.

Symmetries (Even-Odd Relations)

$$\cos(-x) = \cos x \qquad \cosh(-x) = \cosh x$$
$$\sin(-x) = -\sin x \qquad \sinh(-x) = -\sinh x$$

Values for x = 0

$\cos 0 = 1$	$\cosh 0 = 1$
$\sin 0 = 0$	$\sinh 0 = 0$

Values for x = π/2

$\cos \pi/2 = 0$	$\cosh \pi/2 \approx 2.509$
$\sin \pi/2 = 1$	$\sinh \pi/2 \approx 2.301$
	(these values have no special significance)

Addition Formulas

$$\cos(x + y) = \cos x \cos y - \sin x \sin y \qquad \cosh(x + y) = \cosh x \cosh y + \sinh x \sinh y$$

$$\sin(x + y) = \sin x \cos y + \cos x \sin y \qquad \sinh(x + y) = \sinh x \cosh y + \cosh x \sinh y$$

Differentiation Formulas

$$\frac{d}{dx}(\cos x) = -\sin x \qquad \frac{d}{dx}(\cosh x) = \sinh x$$

$$\frac{d}{dx}(\sin x) = \cos x \qquad \frac{d}{dx}(\sinh x) = \cosh x$$

Integration Formulas

$$\int \frac{dx}{\sqrt{1 - x^2}} = \sin^{-1} x + c \qquad \int \frac{dx}{\sqrt{1 + x^2}} = \sinh^{-1} x + c$$

Here $\sin^{-1} x$ and $\sinh^{-1} x$ are the inverse functions of $\sin x$ and $\sinh x$, respectively.

Periodicity

$\cos(x + 2\pi) = \cos x$	no real period
$\sin(x + 2\pi) = \sin x$	

Additional analogies exist between the functions $\tan x$ (defined as $\sin x/\cos x$) and $\tanh x$ (= $\sinh x/\cosh x$) and between the remaining three trigonometric functions $\sec x$ (= $1/\cos x$), $\csc x$ (= $1/\sin x$), and $\cot x$ (= $1/\tan x$) and their hyperbolic counterparts.

It is the periodicity that makes the trigonometric functions so important in mathematics and science. The hyperbolic functions do not have this property and consequently play a less important role; but they are still useful in describing various relations among functions, particularly certain classes of indefinite integrals (antiderivatives).

Interestingly, although the parameter φ in the hyperbolic functions is not an angle, it can be interpreted as twice the area of the hyperbolic sector *OPR* in figure 67, in complete analogy with the interpetation of φ as twice the area of the circular sector *OPR* in figure 66. A proof of this fact—first noted by Vincenzo Riccati around 1750—is given in Appendix 7.

Some Interesting Formulas Involving e

$$e = 1 + \frac{1}{1!} + \frac{1}{2!} + \frac{1}{3!} + \frac{1}{4!} + \ldots$$

This infinite series was discovered by Newton in 1665; it can be obtained from the binomial expansion of $(1 + 1/n)^n$ by letting $n \to \infty$. It converges very quickly, due to the rapidly increasing values of the factorials in the denominators. For example, the sum of the first eleven terms (ending with $1/10!$) is 2.718281801; the true value, rounded to nine decimal places, is 2.718281828.

$$e^{\pi i} + 1 = 0$$

This is Euler's formula, one of the most famous in all of mathematics. It connects the five fundamental constants of mathematics, 0, 1, e, π, and $i = \sqrt{-1}$.

$$e = 2 + \cfrac{1}{1 + \cfrac{1}{2 + \cfrac{2}{3 + \cfrac{3}{4 + \cfrac{4}{5 + \ldots}}}}}$$

This infinite *continued fraction*, and many others involving e and π, was discovered by Euler in 1737. He proved that every rational number can be written as a finite continued fraction, and conversely (the converse is obvious). Hence an infinite (that is, nonterminating) continued fraction always represents an irrational number. Another of Euler's infinite continued fractions involving e is:

$$\frac{e+1}{e-1} = 2 + \cfrac{1}{6 + \cfrac{1}{10 + \cfrac{1}{14 + \ldots}}}$$

$$2 = \frac{e^1}{e^{1/2}} \cdot \frac{e^{1/3}}{e^{1/4}} \cdot \frac{e^{1/5}}{e^{1/6}} \cdot \ldots$$

This *infinite product* can be obtained from the series $\ln 2 = 1 - 1/2 + 1/3 - 1/4 + - \ldots$. It is reminiscent of Wallis's product, $\pi/2 = (2/1) \cdot (2/3) \cdot (4/3) \cdot (4/5) \cdot (6/5) \cdot (6/7) \cdot \ldots$, except that e appears *inside* the product.

✧   ✧   ✧

Applied mathematics abounds in formulas involving e. Here are some examples:

$$\int_0^\infty e^{-x^2/2}\,dx = \sqrt{\frac{\pi}{2}}$$

This definite integral appears in the theory of probability. The *indefinite* integral (antiderivative) of $e^{-x^2/2}$ cannot be expressed in terms of the elementary functions (polynomials and ratios of polynomials, trigonometric and exponential functions, and their inverses); that is, no finite combination of the elementary functions exists whose derivative is $e^{-x^2/2}$.

Another expression whose antiderivative cannot be expressed in terms of the elementary functions is the simple-looking function e^{-x}/x. In fact, its integral, computed from some given x to infinity, *defines* a new function, known as the *exponential integral* and denoted by $Ei(x)$:

$$Ei(x) = \int_x^\infty \frac{e^{-t}}{t}\,dt$$

(the variable of integration is denoted by t so that it will not be confused with the lower limit of integration x). This so-called special function, though not expressible in closed form in terms of the elementary functions, should nevertheless be regarded as known, in the sense that its value for any given positive x has been calculated and tabulated (this is because we can express the integrand e^{-x}/x as a power series and then integrate term by term).

The definite integral $\int_0^\infty e^{-st} f(t)\,dt$ for a given function $f(t)$ has a value that still depends on the parameter s; hence, this integral defines a function $F(s)$ of s, known as the *Laplace transform* of $f(t)$ and written $\mathcal{L}\{f(t)\}$:

$$\mathcal{L}\{f(t)\} = \int_0^\infty e^{-st} f(t)\,dt$$

Because the Laplace transform enjoys many convenient features—all owing to the properties of e^{-st}—it is widely used in applications, particularly in solving linear differential equations (see any text on ordinary differential equations).

13

---

e^{ix}: "The Most Famous of All Formulas"

There is a famous formula—perhaps the most compact and
famous of all formulas—developed by Euler from a
discovery of De Moivre: $e^{i\pi} + 1 = 0$. . . . It appeals
equally to the mystic, the scientist, the philosopher,
the mathematician.

—EDWARD KASNER AND JAMES NEWMAN, *Mathematics*
and the Imagination (1940)

If we compared the Bernoullis to the Bach family, then Leonhard
Euler (1707–1783) is unquestionably the Mozart of mathematics, a
man whose immense output—not yet published in full—is estimated
to fill at least seventy volumes. Euler left hardly an area of mathematics untouched, putting his mark on such diverse fields as analysis,
number theory, mechanics and hydrodynamics, cartography, topology, and the theory of lunar motion. With the possible exception of
Newton, Euler's name appears more often than any other throughout
classical mathematics. Moreover, we owe to Euler many of the mathematical symbols in use today, among them i, π, e and $f(x)$. And as if
that were not enough, he was a great popularizer of science, leaving
volumes of correspondence on every aspect of science, philosophy,
religion, and public affairs.

Leonhard Euler was born in Basel in 1707 to a minister who intended the same career for his son. But Paul Euler was also versed in
mathematics, a subject he had studied under Jakob Bernoulli, and
when he recognized his son's mathematical talents, he changed his
mind. The Bernoullis had something to do with it. Jakob's brother
Johann privately tutored the young Euler in mathematics, and he convinced Paul to let his son pursue his interests. In 1720 Leonhard en-
tered the University of Basel, from which he graduated in just two
years. From then until his death at the age of seventy-six, his mathematical creativity knew no bounds.

His career took him abroad for extended periods. In 1727 he accepted an invitation to join the St. Petersburg Academy of Sciences.

Again the Bernoullis were involved. While receiving lessons from Johann, Euler had befriended his two sons, Daniel and Nicolaus. The young Bernoullis had joined the St. Petersburg Academy some years earlier (tragically, Nicolaus drowned there, prematurely ending the promising career of yet another Bernoulli), and they persuaded the Academy to extend the invitation to Euler. But on the very day that Euler arrived in St. Petersburg to assume his new post, Empress Catherine I died, plunging Russia into a period of uncertainty and repression. The Academy was regarded as an unnecessary drain on the state's budget, and its funds were cut. So Euler began his service there as an adjunct of physiology. Not until 1733 was he given a full professorship in mathematics, succeeding Daniel Bernoulli, who had returned to Basel. In that year, too, Euler married Catherine Gsell; they had thirteen children, but only five survived childhood.

Euler stayed in Russia fourteen years. In 1741 he accepted an invitation by Frederick the Great to join the Berlin Academy of Sciences, as part of the monarch's efforts to attain for Prussia a prominent role in the arts and sciences. Euler stayed there twenty-five years, though not always on good terms with Frederick. The two differed on matters of academic policy as well as in character, the monarch having preferred a more flamboyant person over the quiet Euler. During this period Euler wrote a popular work, *Letters to a German Princess on Diverse Subjects in Physics and Philosophy* (published in three volumes between 1768 and 1772), in which he expressed his views on a wide range of scientific topics (the princess was Frederick's niece, to whom Euler gave private lessons). The *Letters* went through numerous editions and translations. In his entire scientific output—whether technical or expository—Euler always used clear, simple language, making it easy to follow his line of thought.

In 1766 Euler, now nearly sixty years old, accepted an invitation from the new Russian ruler, Catherine II (the "Great"), to return to St. Petersburg (his successor in Berlin was Lagrange). Although the empress bestowed on Euler every possible material benefit, his life during that period was marred by many tragedies. During his first stay in Russia, he had lost the sight in his right eye (according to one account, because of overwork; according to another, because he observed the sun without protecting his eyes). In 1771, during his second stay, he lost the other eye as well. In the same year his house burned down, and many of his manuscripts were lost. Five years later his wife died, but the irrepressible Euler, at age seventy, married again. By now completely blind, he continued his work as before, dictating his numerous results to his children and students. In this he was aided by his phenomenal memory. It is said that he could calculate in his mind with numbers of fifty digits, and he could mentally remember a long sequence of mathematical arguments without writing them on paper. He had enormous powers of concentration and often worked on a difficult problem while his children were sitting on his lap. On 18 September 1783 he was calculating the orbit of the newly discovered planet Uranus. In the evening, while playing with his grandchild, he suddenly had a stroke and died instantly.

It is nearly impossible to do justice to Euler's immense output in this short survey. The enormous range of his work can best be judged from the fact that he founded two areas of research on opposite extremes of the mathematical spectrum: one is number theory, the "purest" of all branches of mathematics; the other is analytical mechanics, the most "applied" of classical mathematics. The former field, despite Fermat's great contributions, was still regarded in Euler's time as a kind of mathematical recreation; Euler made it one of the most respectable areas of mathematical research. In mechanics he reformulated Newton's three laws of motion as a set of differential equations, thus making dynamics a part of mathematical analysis. He also formulated the basic laws of fluid mechanics; the equations governing the motion of a fluid, known as the Euler equations, are the foundation of this branch of mathematical physics. Euler is also regarded as one of the founders of topology (then known as *analysis situs*—"the analysis of position"), the branch of mathematics that deals with continuous deformations of shapes. He discovered the famous formula $V - E + F = 2$ connecting the number of veritces, the number of edges, and the number of faces of any simple polyhedron (a solid having no holes).

The most influential of Euler's numerous works was his *Introductio in analysin infinitorum*, a two-volume work published in 1748 and regarded as the foundation of modern mathematical analysis. In this work Euler summarized his numerous discoveries on infinite series, infinite products, and continued fractions. Among these is the summation of the series $1/1^k + 1/2^k + 1/3^k + \ldots$ for all even values of k from 2 to 26 (for $k = 2$, the series converges to $\pi^2/6$, as Euler had already found in 1736, solving a mystery that had eluded even the Bernoulli brothers). In the *Introductio* Euler made the function the central concept of analysis. His definition of a function is essentially the one we use today in applied mathematics and physics (although in pure mathematics it has been replaced by the "mapping" concept): "A function of a variable quantity is any analytic expression whatsoever made up from that variable quantity and from numbers or constant quantities." The function concept, of course, did not originate with Euler, and Johann Bernoulli defined it in terms very similar to Euler's. But it was Euler who introduced the modern notation $f(x)$ for a function and used it for all kinds of functions—explicit and implicit (in the former the independent variable is isolated on one side of the equation, as in $y = x^2$; in the latter the two variables appear together, as in $2x + 3y = 4$), continuous and discontinuous (his discontinuous functions were actually functions with a discontinuous derivative—a sudden break in the slope of the graph but not in the graph itself), and functions of several independent variables, $u = f(x, y)$ and $u = f(x, y, z)$. And he made free use of the expansion of functions in infinite series and products—often with a carefree attitude that would not be tolerated today.

The *Introductio* for the first time called attention to the central role of the number e and the function e^x in analysis. As already mentioned, until Euler's time the exponential function was regarded merely as the logarithmic function in reverse. Euler put the two functions on an equal basis, giving them independent definitions:

$$e^x = \lim_{n \to \infty} (1 + x/n)^n \tag{1}$$

$$\ln x = \lim_{n \to \infty} n(x^{1/n} - 1). \tag{2}$$

A clue that the two expressions are indeed inverses is this: if we solve the expression $y = (1 + x/n)^n$ for x, we get $x = n(y^{1/n} - 1)$. The more difficult task, apart from interchanging the letters x and y, is to show that the *limits* of the two expressions as $n \to \infty$ define inverse functions. This requires some subtle arguments regarding the limit process, but in Euler's time the nonchalant manipulation of infinite processes was still an accepted practice. Thus, for example, he used the letter i to indicate "an infinite number" and actually wrote the right side of equation 1 as $(1 + x/i)^i$, something that no first-year student would dare today. More's the pity.

$$\left(1 + \frac{x}{\infty} \right)^{\infty}$$

Euler had already used the letter e to represent the number 2.71828 . . . in one of his earliest works, a manuscript entitled "Meditation upon Experiments made recently on the firing of Cannon," written in 1727 when he was only twenty years old (it was not published until 1862, eighty years after his death).[1] In a letter written in 1731 the number e appeared again in connection with a certain differential equation; Euler defines it as "that number whose hyperbolic logarithm is = 1." The earliest appearance of e in a *published* work was in Euler's *Mechanica* (1736), in which he laid the foundations of analytical mechanics. Why did he choose the letter e? There is no general consensus. According to one view, Euler chose it because it is the first letter of the word *exponential*. More likely, the choice came to him naturally as the first "unused" letter of the alphabet, since the letters a, b, c, and d frequently appear elsewhere in mathematics. It seems unlikely that Euler chose the letter because it is the initial of his own name, as has occasionally been suggested: he was an extremely modest man and often delayed publication of his own work so that a colleague or student of his would get due credit. In any event, his choice of the symbol e, like so many other symbols of his, became universally accepted.

Euler used his definition of the exponential function (equation 1) to develop it in an infinite power series. As we saw in Chapter 4, for $x = 1$ equation 1 gives the numerical series

$$\lim_{n \to \infty}\left(1 + \frac{1}{n}\right)^n = 1 + \frac{1}{1!} + \frac{1}{2!} + \frac{1}{3!} + \cdots \tag{3}$$

If we repeat the steps leading to equation 3 (see p. 35) with x/n replacing $1/n$, we get, after a slight manipulation, the infinite series

$$e^x = \lim_{n \to \infty}\left(1 + \frac{x}{n}\right)^n = 1 + \frac{x}{1!} + \frac{x^2}{2!} + \frac{x^3}{3!} + \cdots \tag{4}$$

which is the familiar power series for e^x. It can be shown that this series converges for all real values of x; in fact, the rapidly increasing denominators cause the series to converge very quickly. It is from this series that the numerical values of e^x are usually obtained; the first few terms usually suffice to attain the desired accuracy.

In the *Introductio* Euler also dealt with another kind of infinite process: continued fractions. Take, for example, the fraction 13/8. We can write it as $1 + 5/8 = 1 + 1/(8/5) = 1 + 1/(1 + 3/5)$; that is,

$$\frac{13}{8} = 1 + \frac{1}{1 + \frac{3}{5}}.$$

Euler proved that every rational number can be written as a *finite* continued fraction, whereas an irrational number is represented by an infinite continued fraction, where the chain of fractions never ends. For the irrational number $\sqrt{2}$, for example, we have

$$\sqrt{2} = 1 + \cfrac{1}{2 + \cfrac{1}{2 + \cfrac{1}{2 + \cdots}}}$$

Euler also showed how to write an infinite series as an infinite continued fraction, and vice versa. Thus, using equation 3 as his point of departure, he derived many interesting continued fractions involving the number e, two of which are:

$$e = 2 + \cfrac{1}{1 + \cfrac{1}{2 + \cfrac{2}{3 + \cfrac{3}{4 + \cfrac{4}{5 + \cdots}}}}}$$

$$\sqrt{e} = 1 + \cfrac{1}{1 + \cfrac{1}{1 + \cfrac{1}{1 + \cfrac{1}{5 + \cfrac{1}{1 + \cfrac{1}{1 + \cfrac{1}{1 + \cfrac{1}{9 + \cfrac{1}{1 + \ldots}}}}}}}}}$$

(The pattern in the first formula becomes clear if we move the leading 2 to the left side of the equation; this gives us an expression for the fractional part of e, 0.718281. . . .) These expressions are striking in their regularity, in contrast to the seemingly random distribution of digits in the decimal expansion of irrational numbers.

    Euler was a great experimental mathematician. He played with formulas like a child playing with toys, making all kinds of substitutions until he got something interesting. Often the results were sensational. He took equation 4, the infinite series for e^x, and boldly replaced in it the real variable x with the imaginary expression ix, where $i = \sqrt{-1}$. Now this is the supreme act of mathematical *chutzpah*, for in all our definitions of the function e^x, the variable x has always represented a real number. To replace it with an imaginary number is to play with meaningless symbols, but Euler had enough faith in his formulas to make the meaningless meaningful. By formally replacing x with ix in equation 4, we get

$$e^{ix} = 1 + ix + \frac{(ix)^2}{2!} + \frac{(ix)^3}{3!} + \ldots \tag{5}$$

Now the symbol i, defined as the square root of -1, has the property that its integral powers repeat themselves in cycles of four: $i = \sqrt{-1}$, $i^2 = -1$, $i^3 = -i$, $i^4 = 1$, and so on. Therefore we can write equation 5 as

$$e^{ix} = 1 + ix - \frac{x^2}{2!} - \frac{ix^3}{3!} + \frac{x^4}{4!} + - \ldots . \tag{6}$$

Euler now committed a second sin: he changed the order of terms in equation 6, collecting all the real terms separately from the imaginary terms. This can be dangerous: unlike finite sums, where one can always change the order of terms without affecting the sum, to do so with an infinite series may affect its sum, or even change the series from convergent to divergent.[2] But in Euler's time all this was not yet fully recognized; he lived in an era of carefree experimentation with infinite processes—in the spirit of Newton's fluxions and Leibniz's differentials. Thus, by changing the order of terms in equation 6, he arrived at the series

$$e^{ix} = \left(1 - \frac{x^2}{2!} + \frac{x^4}{4!} - + \ldots\right) + i\left(x - \frac{x^3}{3!} + \frac{x^5}{5!} - + \ldots\right). \tag{7}$$

Now it was already known in Euler's time that the two series appearing in the parentheses are the power series of the trigonometric functions $\cos x$ and $\sin x$, respectively. Thus Euler arrived at the remarkable formula

$$e^{ix} = \cos x + i \sin x, \tag{8}$$

which at once links the exponential function (albeit of an imaginary variable) to ordinary trigonometry.[3] Replacing ix by $-ix$ in equation 8 and using the identities $\cos(-x) = \cos x$ and $\sin(-x) = -\sin x$, Euler obtained the companion equation

$$e^{-ix} = \cos x - i \sin x. \tag{9}$$

Finally, adding and subtracting equations 8 and 9 allowed him to express $\cos x$ and $\sin x$ in terms of the exponential functions e^{ix} and e^{-ix}:

$$\cos x = \frac{e^{ix} + e^{-ix}}{2}, \quad \sin x = \frac{e^{ix} - e^{-ix}}{2i}. \tag{10}$$

These relations are known as the Euler formulas for the trigonometric functions (so many formulas are named after him that just to say "Euler's formula" is not enough).

Although Euler derived many of his results in a nonrigorous manner, each of the formulas mentioned here has withstood the test of rigor—in fact, their proper derivation is today a routine exercise in an advanced calculus class.[4] Euler, like Newton and Leibniz half a century before him, was the pathfinder. The "cleaning up"—the exact, rigorous proof of the numerous results that these three men discovered—was left to a new generation of mathematicans, notably Jean-le-Rond D'Alembert (1717–1783), Joseph Louis Lagrange (1736–1813), and Augustin Louis Cauchy (1789–1857). These efforts continued well into the twentieth century.[5]

The discovery of the remarkable connection between the exponential and the trigonometric functions made it almost inevitable that other unexpected relations would emerge. Thus, by putting $x = \pi$ in equation 8 and knowing that $\cos \pi = -1$ and $\sin \pi = 0$, Euler obtained the formula

$$e^{\pi i} = -1. \tag{11}$$

If "remarkable" is the appropriate description of equations 8 and 9, then one must search for an adequate word to describe equation 11; it must surely rank among the most beautiful formulas in all of mathematics. Indeed, by rewriting it as $e^{\pi i} + 1 = 0$, we obtain a formula that connects the five most important constants of mathematics (and also the three most important mathematical operations—addition, multiplication, and exponentiation). These five constants symbolize the four major branches of classical mathematics: arithmetic, represented by 0 and 1; algebra, by i; geometry, by π; and analysis, by e. No wonder that many people have found in Euler's formula all kinds of mystic meanings. Edward Kasner and James Newman relate one episode in *Mathematics and the Imagination*:

> To Benjamin Peirce, one of Harvard's leading mathematicians in the nineteenth century, Euler's formula $e^{\pi i} = -1$ came as something of a revelation. Having discovered it one day, he turned to his students and said: "Gentlemen, that is surely true, it is absolutely paradoxical; we cannot understand it, and we don't know what it means. But we have proved it, and therefore we know it must be the truth."[6]

NOTES AND SOURCES

1. David Eugene Smith, *A Source Book in Mathematics* (1929; rpt. New York: Dover, 1959), p. 95.

2. For more details, see my book *To Infinity and Beyond: A Cultural History of the Infinite* (1987; rpt. Princeton: Princeton University Press, 1991), pp. 29–39.

3. Euler, however, was not the first to arrive at this formula. Around 1710 the English mathematician Roger Cotes (1682–1716), who helped Newton edit the second edition of the *Principia*, stated the formula log $(\cos\varphi + i\sin\varphi) = i\varphi$, which is equivalent to Euler's formula. This appeared in Cotes's main work, *Harmonia mensurarum*, published posthumously in 1722. Abraham De Moivre (1667–1754), whose name is mentioned in the epigraph to this chapter, discovered the famous formula $(\cos\varphi + i\sin\varphi)^n = \cos n\varphi + i\sin n\varphi$, which in light of Euler's formula becomes the identity $(e^{i\varphi})^n = e^{in\varphi}$. De Moivre was born in France but lived most of his life in London; like Cotes, he was a member of Newton's circle and served on the Royal Society commission that investigated the priority dispute between Newton and Leibniz over the invention of the calculus.

4. To be sure, Euler had his share of blunders. For example, by taking the identity $x/(1-x) + x/(x-1) = 0$ and using long division for each term, he arrived at the formula $\ldots + 1/x^2 + 1/x + 1 + x + x^2 + \ldots = 0$, clearly an absurd result. (Since the series $1 + 1/x + 1/x^2 + \ldots$ converges only for $|x| > 1$, while the series $x + x^2 + \ldots$ converges only for $|x| < 1$, it is meaningless to add the two series.) Euler's carelessness stemmed from the fact that he con-

sidered the value of an infinite series to be the value of the function represented by the series. Today we know that such an interpretation is valid only within the interval of convergence of the series. See Morris Kline, *Mathematics: The Loss of Certainty* (New York: Oxford University Press, 1980), pp. 140–145.

5. Ibid., ch. 6.

6. (New York: Simon and Schuster, 1940), pp. 103–104. Peirce's admiration of Euler's formula led him to propose two rather unusual symbols for π and e (see p. 162).

A Curious Episode in the History of e

Benjamin Peirce (1809–1880) became professor of mathematics at Harvard College at the young age of twenty-four.[1] Inspired by Euler's formula $e^{\pi i} = -1$, he devised new symbols for π and e, reasoning that

The symbols which are now used to denote the Naperian base and the ratio of the circumference of a circle to its diameter are, for many reasons, inconvenient; and the close relation between these two quantities ought to be indicated in their notation. I would propose the following characters, which I have used with success in my lectures: —

    ⋒ to denote ratio of circumference to diameter,

    ⋒ to denote Naperian base.

It will be seen that the former symbol is a modification of the letter c (*circumference*), and the latter of b (*base*). The connection of these quantities is shown by the equation,

$$⋒^⋒ = (-1)^{-\sqrt{-1}}$$

Peirce published his suggestion in the *Mathematical Monthly* of February 1859 and used it in his book *Analytic Mechanics* (1855). His two sons, Charles Saunders Peirce and James Mills Peirce, also mathematicians, continued to use their father's notation, and James Mills decorated his *Three and Four Place Tables* (1871) with the equation $\sqrt{e^{\pi}} = \sqrt[i]{i}$ (fig. 68).[2]

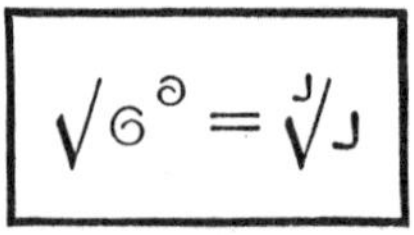

$$\boxed{\sqrt{6^{\partial}} = \sqrt[J]{J}}$$

FIG. 68. Benjamin Peirce's symbols for π, e, and i appear on the title page of James Mills Peirce's *Three and Four Place Tables* (Boston, 1871). The formula is Euler's $e^{\pi i} = -1$ in disguise. Reprinted from Florian Cajori, *A History of Mathematical Notations* (1928–1929; La Salle, Ill.: Open Court, 1951), with permission.

Not surprisingly, Peirce's suggestion was not received with great enthusiasm. Aside from the typographical difficulties of printing his symbols, it takes a bit of skill to distinguish his ⋒ from his ⋒. His students, we are told, preferred the more traditional π and e.[3]

Notes and Sources

1. David Eugene Smith, *History of Mathematics*, 2 vols. (1923; rpt. New York: Dover, 1958), 1:532.

2. This equation, as well as Benjamin Peirce's equation $e^{\pi} = (-1)^{-i}$, can be derived from Euler's formula by a formal manipulation of the symbols.

3. Florian Cajori, *A History of Mathematical Notations*, vol. 2, *Higher Mathematics* (1929; rpt. La Salle, Ill.: Open Court, 1929), pp. 14–15.

14

---

e^{x+iy}: The Imaginary Becomes Real

That this subject [imaginary numbers] has hitherto been

surrounded by mysterious obscurity, is to be attributed

largely to an ill-adapted notation. If, for instance, +1, −1,

√−1 had been called direct, inverse, and lateral units,

instead of positive, negative, and imaginary (or even

impossible), such an obscurity would have been

out of the question.

—CARL FRIEDRICH GAUSS (1777–1855)[1]

The introduction of expressions like e^{ix} into mathematics raises the question: What, exactly, do we mean by such an expression? Since the exponent is imaginary, we cannot calculate the values of e^{ix} in the same sense that we can find the value of, say, $e^{3.52}$—unless, of course, we clarify what we mean by "calculate" in the case of imaginary numbers. This takes us back to the sixteenth century, when the quantity $\sqrt{-1}$ first appeared on the mathematical scene.

An aura of mysticism still surrounds the concept that has since been called "imaginary numbers," and anyone who encounters these numbers for the first time is intrigued by their strange properties. But "strange" is relative: with sufficient familiarity, the strange object of yesterday becomes the common thing of today. From a mathematical point of view, imaginary numbers are no more strange than, say, negative numbers; they are certainly simpler to handle than ordinary fractions, with their "strange" law of addition $a/b + c/d = (ad + bc)/bd$. Indeed, of the five famous numbers that appear in Euler's formula $e^{\pi i} + 1 = 0$, $i = \sqrt{-1}$ is perhaps the least interesting. It is the *consequences* of accepting this number into our number system that make imaginary numbers—and their extension to complex numbers—so important in mathematics.

Just as negative numbers arose from the need to solve the linear equation $x + a = 0$ when a is positive, so did imaginary numbers arise from the need to solve the quadratic equation $x^2 + a = 0$ when a is positive. Specifically, the number $\sqrt{-1}$, the "imaginary unit," is defined as one of the two solutions of the equation $x^2 + 1 = 0$ (the other being $-\sqrt{-1}$), just as the number -1, the "negative unit," is defined as the solution of the equation $x + 1 = 0$. Now, to solve the equation $x^2 + 1 = 0$ means to find a number whose square is -1. Of course, no real number will do, because the square of a real number is never negative. Thus *in the domain of real numbers* the equation $x^2 + 1 = 0$ has no solutions, just as in the domain of positive numbers the equation $x + 1 = 0$ has no solution.

For two thousand years mathematics thrived without bothering about by these limitations. The Greeks (with one known exception: Diophantus in his *Arithmetica*, ca. 275 A.D.) did not recognize negative numbers and did not need them; their main interest was in geometry, in quantities such as length, area, and volume, for the description of which positive numbers are entirely sufficient. The Hindu mathematician Brahmagupta (ca. 628) used negative numbers, but medieval Europe mostly ignored them, regarding them as "imaginary" or "absurd." Indeed, so long as one regards subtraction as an act of "taking away," negative numbers are absurd: one cannot take away, say, five apples from three. Negative numbers, however, kept forcing themselves upon mathematics in other ways, mainly as roots of quadratic and cubic equations but also in connection with practical problems (Leonardo Fibonacci, in 1225, interpreted a negative root arising in a financial problem as a loss instead of a gain). Still, even during the Renaissance, mathematicians felt uneasy about them. An important step toward their ultimate acceptance was taken by Rafael Bombelli (born ca. 1530), who interpreted numbers as lengths on a line and the four basic arithmetic operations as movements along the line, thus giving a geometric interpretation to real numbers. But only when it was realized that subtraction could be interpreted as the *inverse of addition* was a full acceptance of negative numbers into our number system made possible.[2]

Imaginary numbers have undergone a similar evolution. The impossibility of solving the equation $x^2 + a = 0$ when a is positive had been known for centuries, but attempts to overcome the difficulty were slow in coming. One of the first was made in 1545 when the Italian Girolamo Cardano (1501–1576) tried to find two numbers whose sum is 10 and whose product is 40. This leads to the quadratic equation $x^2 - 10x + 40 = 0$, whose two solutions—easily found from the quadratic formula—are $5 + \sqrt{-15}$ and $5 - \sqrt{-15}$. At first Cardano did not know what to do with these "solutions" because he could not find their values. But he was intrigued by the fact that if he operated on these imaginary solutions in a purely formal way, as if they obeyed all the rules of ordinary arithmetic, the two solutions indeed fulfilled the conditions of the problem: $(5 + \sqrt{-15}) + (5 - \sqrt{-15}) = 10$

and $(5 + \sqrt{-15}) \cdot (5 - \sqrt{-15}) = 25 - 5\sqrt{-15} + 5\sqrt{-15} - (\sqrt{-15})^2 = 25 - (-15) = 40$.

With the passage of time, quantities of the form $x + (\sqrt{-1})y$—nowadays called *complex numbers* and written as $x + iy$, where x and y are real numbers and $i = \sqrt{-1}$—increasingly found their way into mathematics. For example, the solution of the general cubic (third-degree) equation requires one to deal with these quantities, even if the final solutions turn out to be real. It was not until the beginning of the nineteenth century, however, that mathematicians felt comfortable enough with complex numbers to accept them as bona fide numbers.

Two developments greatly helped in this process. First, around 1800, it was shown that the quantity $x + iy$ could be given a simple geometric interpretation. In a rectangular coordinate system we plot the point P whose coordinates are x and y. If we interpret the x and y axes as the "real" and "imaginary" axes, respectively, then the complex number $x + iy$ is represented by the point $P(x, y)$, or equivalently by the line segment (vector) OP (fig. 69). We can then add and subtract complex numbers in the same way that we add and subtract vectors, by separately adding or subtracting the real and imaginary components: for example, $(1 + 3i) + (2 - 5i) = 3 - 2i$ (fig. 70). This graphic representation was suggested at about the same time by three scientists in different countries: Caspar Wessel (1745–1818), a Norwegian surveyor, in 1797; Jean Robert Argand (1768–1822) of France in 1806; and Carl Friedrich Gauss (1777–1855) of Germany in 1831.

The second development was due to the Irish mathematician Sir William Rowan Hamilton (1805–1865). In 1835 he defined complex numbers in a purely formal way by treating them as *ordered pairs* of real numbers subject to certain rules of operation. A "complex number" is defined as the ordered pair (a, b), where a and b are real numbers. Two pairs (a, b) and (c, d) are equal if and only if $a = c$ and $b = d$. Multiplying the pair (a, b) by a real number k (a "scalar") produces the pair (ka, kb). The sum of the pairs (a, b) and (c, d) is the pair $(a + c, b + d)$ and their product is the pair $(ac - bd, ad + bc)$. The meaning behind the seemingly strange definition of multiplication becomes clear if we multiply the pair $(0, 1)$ by itself: according to the rule just given, we have $(0, 1) \cdot (0, 1) = (0 \cdot 0 - 1 \cdot 1, 0 \cdot 1 + 1 \cdot 0) = (-1, 0)$. If we now agree to denote any pair whose second component is 0 by the letter denoting its first component and regard it as a "real" number—that is, if we identify the pair $(a, 0)$ with the real number a—then we can write the last result as $(0, 1) \cdot (0, 1) = -1$. Denoting the pair $(0, 1)$ by the letter i, we thus have $i \cdot i = -1$, or simply $i^2 = -1$. Moreover, we can now write any pair (a, b) as $(a, 0) + (0, b) = a(1, 0) + b(0, 1) = a \cdot 1 + b \cdot i = a + ib$, that is, as an ordinary com-

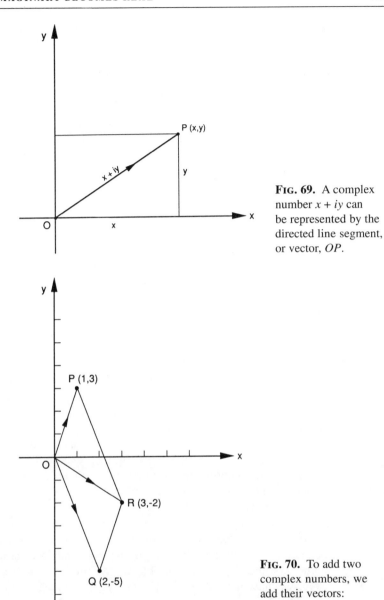

FIG. 69. A complex number $x + iy$ can be represented by the directed line segment, or vector, OP.

FIG. 70. To add two complex numbers, we add their vectors: $(1 + 3i) + (2 - 5i) = 3 - 2i$.

plex number. Thus we have removed from the complex numbers any remaining trace of mystery; indeed, the only reminder of their troublesome evolution is the symbol i for "imaginary." Hamilton's rigorous approach marked the beginning of axiomatic algebra: the step-by-step development of a subject from a small set of simple definitions ("axioms") and a chain of logical consequences ("theorems") derived from them. The axiomatic method was not new to mathematics, of course; it had been dogmatically followed in geometry ever since the Greeks established this science as a rigorous, deductive mathematical discipline, immortalized in Euclid's *Elements* (ca. 300 B.C.). Now, in the mid-1800s, algebra was emulating geometry's example.

Once the psychological difficulty of accepting complex numbers was overcome, the road to new discoveries was open. In 1799, in his doctoral dissertation at the age of twenty-two, Gauss gave the first rigorous demonstration of a fact that had been known for some time: a polynomial of degree n (see p. 98) always has at least one root in the domain of complex numbers (in fact, if we count repeated roots as separate roots, a polynomial of degree n has exactly n complex roots).[3] For example, the polynomial $x^3 - 1$ has the three roots (that is, solutions of the equation $x^3 - 1 = 0$) 1, $(-1 + i\sqrt{3})/2$ and $(-1 - i\sqrt{3})/2$, as can easily be checked by computing the cube of each number. Gauss's theorem is known as the Fundamental Theorem of Algebra; it shows that complex numbers are not only necessary to solve a general polynomial equation, they are also *sufficient*.[4]

The acceptance of complex numbers into the realm of algebra had an impact on analysis as well. The great success of the differential and integral calculus raised the possibility of extending it to *functions of complex variables*. Formally, we can extend Euler's definition of a function (p. 155) to complex variables without changing a single word; we merely allow the constants and variables to assume complex values. But from a geometric point of view, such a function cannot be plotted as a graph in a two-dimensional coordinate system because *each* of the variables now requires for its representation a two-dimensional coordinate system, that is, a plane. To interpret such a function geometrically, we must think of it as a *mapping*, or transformation, from one plane to another.

Let us illustrate this with the function $w = z^2$, where both z and w are complex variables. To describe this function geometrically, we need two coordinate systems, one for the independent variable z and another for the dependent variable w. Writing $z = x + iy$ and $w = u + iv$, we have $u + iv = (x + iy)^2 = (x + iy)(x + iy) = x^2 + xiy + iyx + i^2y^2 = x^2 + 2ixy - y^2 = (x^2 - y^2) + i(2xy)$. Equating the real and imaginary parts on both sides of this equation, we get $u = x^2 - y^2$, $v = 2xy$. Now suppose that we allow the variables x and y to trace some curve in the "z-plane" (the xy plane). This will force the variables u and v to trace an image curve in the "w-plane" (the uv plane). For example, if the point $P(x, y)$ moves along the hyperbola $x^2 - y^2 = c$ (where c is a constant), the image point $Q(u, v)$ will move along the curve $u = c$, that is, along a vertical line in the w-plane. Similarly, if P moves along the hyperbola $2xy = k =$ constant, Q will trace the horizontal

[handwritten left margin:] Crucial definition or underlying mechanism.

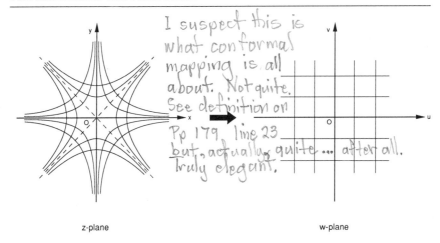

z-plane   *w-plane*

FIG. 71. Mapping by the complex function $w = z^2$.

line $v = k$ (fig. 71). The hyperbolas $x^2 - y^2 = c$ and $2xy = k$ form two families of curves in the z-plane, each curve corresponding to a given value of the constant. Their image curves form a rectangular grid of horizontal and vertical lines in the w-plane.

Can we differentiate a function $w = f(z)$, where both z and w are complex variables, in the same way that we differentiate a function $y = f(x)$ of the real variables x and y? The answer is yes—with a caveat. To begin, we can no longer interpret the derivative of a function as the slope of the tangent line to its graph, because a function of a complex variable cannot be represented by a single graph; it is a mapping from one plane to another. Still, we may try to perform the differentiation process in a purely formal way by finding the difference in the values of $w = f(z)$ between two "neighboring" points z and $z + \Delta z$, dividing this difference by Δz, and going to the limit as $\Delta z \to 0$. This would give us, at least formally, a measure of the rate of change of $f(z)$ at the point z. But even in this formal process we encounter a difficulty that does not exist for functions of a real variable.

Inherent in the concept of limit is the assumption that the end result of the limiting process is the same, regardless of how the independent variable approaches its "ultimate" value. For example, in finding the derivative of $y = x^2$ (p. 86), we started with some fixed value of x, say x_0, then moved to a neighboring point $x = x_0 + \Delta x$, found the difference Δy in the values of y between these points, divided this difference by Δx, and finally found the limit of $\Delta y / \Delta x$ as $\Delta x \to 0$. This gave us $2x_0$, the value of the derivative at x_0. Now, in letting Δx approach 0, we assumed—though we never said so explicitly—that the same result should be obtained regardless of how we let $\Delta x \to 0$. For instance, we could let Δx approach 0 through positive values only (that is, let x approach x_0 from the right side), or through negative values only (x approaches x_0 from the left). The tacit assumption is that the final result—the derivative of $f(x)$ at x_0—is independent of the manner in which $\Delta x \to 0$. For the great majority of functions we encounter in elementary algebra this is a subtle, almost pedantic detail, because these functions are usually smooth and continuous—their graphs have no sharp corners or sudden breaks. Hence we need not be overly concerned when computing the derivatives of these functions.[5]

When it comes to functions of a complex variable, however, these considerations at once become crucial. Unlike the real variable x, a complex variable z can approach a point z_0 from infinitely many directions (recall that the independent variable alone requires an entire plane for its representation). Thus, to say that the limit of $\Delta w/\Delta z$ as $\Delta z \to 0$ exists implies that the (complex) value of this limit should be independent of the particular direction along which $z \to z_0$.

It can be shown that this formal requirement leads to a pair of differential equations of the utmost importance in the calculus of functions of a complex variable. These are known as the Cauchy-Riemann equations, named for Augustin Louis Cauchy (1789–1857) of France and Georg Friedrich Bernhard Riemann (1826–1866) of Germany. To derive these equations would go beyond the scope of this book,[6] so let show only how they work. Given a function $w = f(z)$ of a complex variable z, if we write $z = x + iy$ and $w = u + iv$, then both u and v become (real-valued) functions of the (real) variables x and y; in symbols, $w = f(z) = u(x, y) + iv(x, y)$. For example, in the case of the function $w = z^2$ we found that $u = x^2 - y^2$ and $v = 2xy$. The Cauchy-Riemann equations say that for a function $w = f(z)$ to be differentiable (that is, to have a derivative) at a point z in the complex plane, the derivative of u with respect to x must equal the derivative of v with respect to y, and the derivative of u with respect to y must equal the *negative* derivative of v with respect to x, all derivatives being evaluated at the point $z = x + iy$ in question.

It would, of course, be much simpler to express these relations in mathematical language instead of words, but we must first introduce a new notation for the derivative in this case. This is because both u and v are functions of *two* independent variables, and we must state with respect to which variable are we differentiating. We denote the derivatives just mentioned by the symbols $\partial u/\partial x$, $\partial u/\partial y$, $\partial v/\partial x$, and $\partial v/\partial y$. The operations $\partial/\partial x$ and $\partial/\partial y$ are called *partial differentiations* with respect to x and y, respectively. In performing these differentiations, we keep fixed all the variables except those indicated by the differentiation symbol. Thus in $\partial/\partial x$ we keep y fixed, while in $\partial/\partial y$ we keep x fixed. The Cauchy-Riemann equations say that

$$\frac{\partial u}{\partial x} = \frac{\partial v}{\partial y}, \quad \frac{\partial u}{\partial y} = -\frac{\partial v}{\partial x}. \tag{1}$$

[margin note: Subtle. The variable we keep fixed becomes momentarily a "constant" and so it "drops out" when we differentiate.]

For the function $w = z^2$, we have $u = x^2 - y^2$ and $v = 2xy$, so that $\partial u/\partial x = 2x$, $\partial u/\partial y = -2y$, $\partial v/\partial x = 2y$, and $\partial v/\partial y = 2x$. The Cauchy-Riemann equations are thus satisfied for all values of x and y, and consequently $w = z^2$ is differentiable at every point z of the complex plane. Indeed, if we formally repeat the process of finding the derivative of $y = x^2$ (see p. 86) with x replaced by z and y by w, we get $dw/dz = 2z$. This formula gives the (complex) value of the derivative for each point in the z-plane. The Cauchy-Riemann equations, although not directly involved in computing the derivative, provide a necessary (and, with a slight change in the assumptions, also sufficient) condition for the derivative to *exist* at the point in question.

[margin note: How about that?!]

If a function $w = f(z)$ is differentiable at a point z of the complex plane, we say that $f(z)$ is *analytic* at z. In order for this to happen, the Cauchy-Riemann equations must be fulfilled there. Thus analyticity is a much stronger requirement than mere differentiability in the real domain. But once a function is shown to be analytic, it obeys all the familiar rules of differentiation that apply to functions of a real variable. For example, the differentiation formulas for the sum and product of two functions, the chain rule, and the formula $d(x^n)/dx = nx^{-1}$ all continue to hold when the real variable x is replaced by the complex variable z. We say that the properties of the function $y = f(x)$ are *carried over to the complex domain*.

After this rather technical excursion into the general theory of complex functions, we are ready to return to our subject: the exponential function. Taking as our point of departure Euler's formula $e^{ix} = \cos x + i \sin x$, we can regard the right side of this equation as the *definition* of the expression e^{ix}, <u>which until now has never been defined.</u> But we can do better than that: having allowed the exponent to assume imaginary values, why not let it assume *complex* values as well? In other words, we wish to give a meaning to the expression e^z when $z = x + iy$. We can try to work our way in a purely manipulative manner, in the spirit of Euler. Assuming that e^z obeys all the familiar rules of the exponential function of a real variable, we have

[margin note: Thus causing me much confusion. eek!]

[interlinear note: by algebraic legerdemain]

$$e^z = e^{x+iy} = e^x e^{iy} = e^x(\cos y + i \sin y). \tag{2}$$

Of course, the weak link in this argument is the very assumption just made—that the undefined expression e^z behaves according to the good old rules of algebra of real variables. It is really an act of faith, and of all the sciences, mathematics is the least forgiving of acts of faith. But there is a way out: why not turn the tables and *define* e^z by equation 2? This we are certainly free to do, for nothing in the defini-

[margin note: yeah!]

[margin note: Good quote for a chapter heading.]

[handwritten note: Sure. Why not? Definition by decree.]

tion will contradict what has already been established about the exponential function.

Of course, in mathematics we are free to define a new object in any way we want, so long as the definition does not contradict any previously accepted definitions or established facts. The real question is: Is the definition justified by the properties of the new object? In our case, the justification for denoting the left side of equation 2 by e^z is the fact that this definition ensures that the new object, the exponential function of a complex variable, behaves exactly as we want it to: it preserves all the formal properties of the real-valued function e^x. For example, just as we have $e^{x+y} = e^x \cdot e^y$ for any two real numbers x and y, so we have $e^{w+z} = e^w \cdot e^z$ for any two complex numbers w and z.[7] Moreover, if z is real (that is, if $y = 0$), the right side of equation 2 gives us $e^x(\cos 0 + i \sin 0) = e^x(1 + i \cdot 0) = e^x$, so that the exponential function of a real variable is included as a special case in the definition of e^z.

What about the derivative of e^z? It can be shown that if a function $w = f(z) = u(x, y) + iv(x, y)$ is differentiable at a point $z = x + iy$, its derivative there is given by

$$\frac{dw}{dz} = \frac{\partial u}{\partial x} + i\frac{\partial v}{\partial x} \tag{3}$$

Whoops!

(or alternatively by $\partial v/\partial y - i\partial u/\partial y$; the two expressions are equal in

Twin differentials light of the Cauchy-Riemann equations). For the function $w = e^z$,

"identical" equation 2 gives $u = e^x \cos y$ and $v = e^x \sin y$, so that $\partial u/\partial x = e^x \cos y$

twins, so to and $\partial v/\partial x = e^x \sin y$. We therefore have

speak.

$$\frac{d}{dz}(e^z) = e^x(\cos y + i \sin y) = e^z. \tag{4}$$

Thus the function e^z is equal to its own derivative, exactly as with the function e^x.

We should mention that there is an alternative approach to developing the theory of functions of a complex variable, or the *theory of*

theory of *functions*, as it is known for short. This approach, pioneered by

functions Cauchy and perfected by the German mathematician Karl Weierstrass (1815–1897), makes extensive use of power series. The function e^z, for example, is defined as

Thank you,

uncle Karl.

$$e^z = 1 + \frac{z}{1!} + \frac{z^2}{2!} + \frac{z^3}{3!} + \cdots, \tag{5}$$

a definition motivated by Euler's definition of e^x as the limit of $(1 + x/n)^n$ when $n \to \infty$ (see p. 157). The details go beyond the scope of this book, but the essense of the argument is to show that the power series (5) converges for all values of z in the complex plane

I think I like this approach better. It is more concrete & computational & less mysterious & magical.

and that it can be differentiated term by term, exactly as with an ordinary (finite) polynomial. All the properties of e^z can then be derived from this definition; in particular, the formula $d(e^z)/dz = e^z$ follows immediately from a term-by-term differentiation of the series (5), as the reader can easily verify.

At this point we have extended the exponential function to the complex domain in such a way that all its familiar properties from the real domain are preserved. But what good does this do? What new information have we gained? Indeed, if it were only a matter of formally replacing the real variable x with the complex variable z, the process would hardly be justified. Luckily, the extension of a function to the complex domain carries with it some real bonuses. We have already seen one of them: the interpretation of a complex function as a mapping from the z-plane to the w-plane.

To see what kind of a mapping is effected by the function $w = e^z$, we must digress briefly from our main subject and talk about the *representation* polar representation of a complex number. As we saw in Chapter 11, we can locate a point P in the plane either by its rectangular coordinates (x, y) or by its polar coordinates (r, θ). From the right triangle *OPR* in figure 72 we see that the two pairs of coordinates are related through the formulas $x = r\cos\theta$, $y = r\sin\theta$. We can therefore write any complex number $z = x + iy$ as $z = r\cos\theta + ir\sin\theta$, or, after factoring out r, $\theta \equiv$ *theta*

$$z = x + iy = r(\cos\theta + i\sin\theta). \tag{6}$$

We can shorten equation 6 even more by replacing the expression $\cos\theta + i\sin\theta$ with the abbreviated symbol $\operatorname{cis}\theta$. We thus have

$$z = x + iy = r\operatorname{cis}\theta. \tag{7}$$

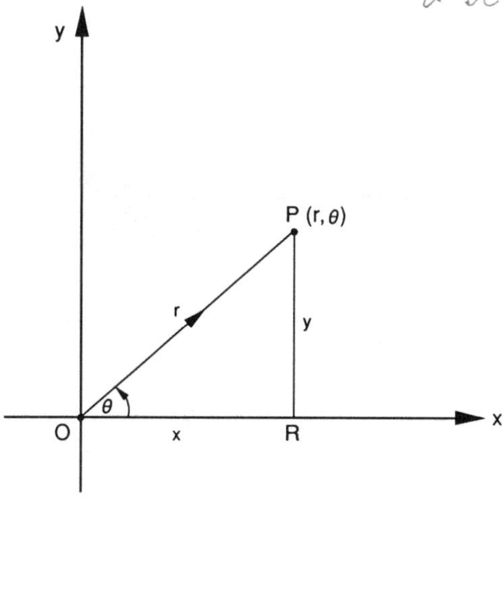

Here comes cis θ & De Moivre

FIG. 72. Polar representation of a complex number.

The two forms of a complex number, $x + iy$ and $r\operatorname{cis}\theta$, are known as the rectangular and polar representations of z, respectively (here, as always in analysis, the angle θ is measured in radians [see p. 121]). As an example, the number $z = 1 + i$ has the polar representation $\sqrt{2}\operatorname{cis}\pi/4$, because the distance of the point $P(1, 1)$ from the origin is $r = \sqrt{(1^2 + 1^2)} = \sqrt{2}$, and the line segment OP forms an angle of $\theta = 45° = \pi/4$ radians with the positive x-axis.

θ measured in radians, not degrees.

$\frac{\pi}{4} = 45° = \frac{1}{8}\theta$

The polar representation turns out to be particularly useful when multiplying or dividing two complex numbers. Let $z_1 = r_1\operatorname{cis}\theta$ and $z_2 = r_2\operatorname{cis}\varphi$. Then $z_1 z_2 = (r_1\operatorname{cis}\theta)(r_2\operatorname{cis}\varphi) = r_1 r_2(\cos\theta + i\sin\theta)$ $(\cos\varphi + i\sin\varphi) = r_1 r_2[(\cos\theta\cos\varphi - \sin\theta\sin\varphi) + i(\cos\theta\sin\varphi + \sin\theta\cos\varphi)]$. If we make use of the addition formulas for sine and cosine (see p. 149), the expressions inside the parentheses become simply $\cos(\theta + \varphi)$ and $\sin(\theta + \varphi)$, so that $z_1 z_2 = r_1 r_2\operatorname{cis}(\theta + \varphi)$. This means that in order to multiply two complex numbers, we must multiply their distances from the origin and *add* their angles. In other words, the distance undergoes a *dilation* (stretching), while the angle undergoes a *rotation*. It is this geometric interpretation that makes complex numbers so useful in numerous applications—from mechanical vibrations to electric circuits—indeed, whenever rotations are involved.

Going back to equation 2, we see that its right side has exactly the form of a polar representation, with e^x playing the role of r and y the role of θ. Thus, if we represent the variable $w = e^z$ in polar form as $R(\cos\Phi + i\sin\Phi)$, we have $R = e^x$ and $\Phi = y$. Now imagine that a point P in the z-plane moves along the horizontal line $y = c = $ constant. Then its image point Q in the w-plane will move along the ray $\Phi = c$ (fig. 73). In particular, the line $y = 0$ (the x-axis) is mapped on

Φ = phi

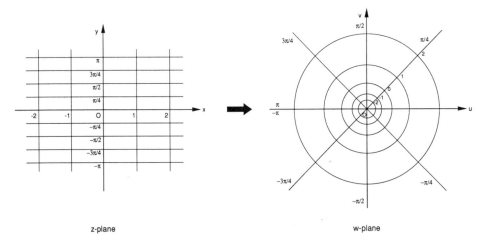

z-plane   w-plane

FIG. 73. Mapping by the complex function $w = e^z$.

the ray $\Phi = 0$ (the positive u-axis), the line $y = \pi/2$, on the ray $\Phi = \pi/2$ (the positive v-axis), the line $y = \pi$, on the ray $\Phi = \pi$ (the negative u-axis), and—surprise!—the line $y = 2\pi$ is mapped again on the positive u-axis. This is because the functions $\sin y$ and $\cos y$ appearing in equation 2 are periodic—their values repeat every 2π radians (360°). But this means that the function e^z itself is periodic—indeed, it has an *imaginary period* of $2\pi i$. And just as it is sufficient to know the behavior of the real-valued functions $\sin x$ and $\cos x$ within a single period, say from $x = -\pi$ to $x = \pi$, so it is sufficient to know the behavior of the complex-valued function e^z in a single horizontal strip, say from $y = -\pi$ to $y = \pi$ (more precisely, $-\pi < y \le \pi$), called the *fundamental domain* of e^z.

So much for horizontal lines. When P moves along the vertical line $x = k = \text{constant}$, its image Q moves along the curve $R = e^k = \text{constant}$, that is, on a circle with center at the origin and radius $R = e^k$ (see again fig. 73). For different vertical lines (different values of k) we get different circles, all concentric to the origin. Note, however, that if the lines are spaced equally, their image circles increase *exponentially*—their radii grow in a geometric progression. We find in this a reminder that the function e^z has its genealogical roots in the famous relation between arithmetic and geometric progressions that had led Napier to invent his logarithms in the early seventeenth century.

✧   ✧   ✧

The inverse of the real-valued function $y = e^x$ is the natural logarithmic function $y = \ln x$. In exactly the same way, the inverse of the complex-valued function $w = e^z$ is the *complex natural logarithm* of z, $w = \ln z$. There is, however, an important difference. The function $y = e^x$ has the property that two different values of x always produce two different values of y; this can be seen from the graph of e^x (Chapter 10, fig. 31), which increases from left to right along the entire x-axis. A function that has this property is said to be *one-to-one*, written 1:1. An example of a function that is *not* 1:1 is the parabola $y = x^2$, because we have, for example, $(-3)^2 = 3^2 = 9$. Strictly speaking, only a 1:1 function has an inverse, because only then will each value of y be the image of exactly one x value. Hence the function $y = x^2$ does not have an inverse (though we can remedy the situation by restricting the domain to $x \ge 0$). For the same reason, the trigonometric functions $y = \sin x$ and $y = \cos x$ have no inverses; the fact that these functions are periodic means that infinitely many x values produce the same y (again, the situation can be remedied by an appropriate restriction of the domain).

We saw earlier that the complex function e^z is periodic. Therefore, if we were to abide by the rules of real-valued functions, this function would not have an inverse. However, because many of the common functions of a real variable become periodic when extended to the complex domain, it is customary to relax the $1:1$ restriction and allow a function of a complex variable to have an inverse even if it is not $1:1$. This means that the inverse function will assign to each value of the independent variable several values of the dependent variable. The complex logarithm is an example of such a *multivalued function*.

Our goal is to express the function $w = \ln z$ in complex form as $u + iv$. We start with $w = e^z$ and express w in polar form as $R\operatorname{cis}\Phi$. By equation 2 we then have $R\operatorname{cis}\Phi = e^x\operatorname{cis}y$. Now, two complex numbers are equal only if they have the same distance from the origin and the same direction with respect to the real axis. The first of these conditions gives us $R = e^x$. But the second condition is fulfilled not only when $\Phi = y$ but also when $\Phi = y + 2k\pi$, where k is any integer, positive or negative. This is because a given ray emanating from the origin corresponds to infinitely many angles, differing from one another by any number of full rotations (that is, integral multiples of 2π). We thus have $R = e^x$, $\Phi = y + 2k\pi$. Solving these equations for x and y in terms of R and Φ, we get $x = \ln R$, $y = \Phi + 2k\pi$ (actually $\Phi - 2k\pi$, but the negative sign is irrelevant because k can be any positive or negative integer). We therefore have $z = x + iy = \ln R + i(\Phi + 2k\pi)$. Interchanging as usual the letters for the independent and dependent variables, we finally have

$$w = \ln z = \ln r + i(\theta + 2k\pi), \quad k = 0, \pm 1, \pm 2, \ldots. \tag{8}$$

Equation 8 defines the *complex logarithm* of any complex number $z = r\operatorname{cis}\theta$. As we see, this logarithm is a multivalued function: a given number z has infinitely many logarithms, differing from one another by multiples of $2\pi i$. As an example, let us find the logarithm of $z = 1 + i$. The polar form of this number is $\sqrt{2}\operatorname{cis}\pi/4$, so that $r = \sqrt{2}$ and $\theta = \pi/4$. By equation 8 we have $\ln z = \ln\sqrt{2} + i(\pi/4 + 2k\pi)$. For $k = 0, 1, 2, \ldots$ we get the values $\ln\sqrt{2} + i(\pi/4) \approx 0.3466 + 0.7854i$, $\ln\sqrt{2} + i(9\pi/4) \approx 0.3466 + 7.0686i$, $\ln\sqrt{2} + i(17\pi/4) \approx 0.3466 + 13.3518i$, and so on; additional values are obtained when k is negative.

What about the logarithm of a *real* number? Since the real number x is also the complex number $x + 0i$, we expect that the natural logarithm of $x + 0i$ should be identical with the natural logarithm of x. This indeed is true—almost. The fact that the complex logarithm is a multivalued function introduces additional values not included in the natural logarithm of a real number. Take the number $x = 1$ as an example. We know that $\ln 1 = 0$ (because $e^0 = 1$). But when we regard the real number 1 as the complex number $z = 1 + 0i = 1\operatorname{cis}0$, we have from equation 8 $\ln z = \ln 1 + i(0 + 2k\pi) = 0 + i(2k\pi) = 2k\pi i$,

[Handwritten marginalia:]

24 Sep 1994

If we diagram a 1 handed tensile structure on the z-plane, what does it look like on the w-plane and vice versa

He actually reduces the thing to some numbers.

where $k = 0, \pm1, \pm2, \ldots$. Thus the complex number $1 + 0i$ has infinitely many logarithms—$0, \pm2\pi i, \pm4\pi i$, and so on—all except 0 being purely imaginary. The value 0—and, more generally, the value $\ln r + i\theta$ obtained by letting $k = 0$ in equation 8—is called the *principal value* of the logarithm and denoted by Ln z.

✧ ✧ ✧   *cap L en 2cc*

Let us now return to the eighteenth century and see how these ideas took hold. As we recall, the problem of finding the area under the hyperbola $y = 1/x$ was one of the outstanding mathematical problems of the seventeenth century. The discovery that this area involves logarithms shifted the focus from the original role of logarithms as a computational device to the properties of the logarithmic *function*. It was Euler who gave us the modern definition of logarithm: if $y = b^x$, where b is any positive number different from 1, then $x = \log_b y$ (read *ot..ot..ot..* "logarithm base b of y"). Now, so long as the variable x is real, $y = b^x$ will always be positive; therefore, *in the domain of real numbers* the logarithm of a negative number does not exist, just as the square root of a negative number does not exist in the domain of real numbers. But by the eighteenth century complex numbers were already well integrated into mathematics, so naturally the question arose: What is the logarithm of a negative number? In particular, what is $\ln(-1)$?

This question gave rise to a lively debate. The French mathematician Jean-le-Rond D'Alembert (1717–1783), who died the same year as Euler, thought that $\ln(-x) = \ln x$, and therefore $\ln(-1) = \ln 1 = 0$. His rationale was that since $(-x)(-x) = x^2$, we should have $\ln[(-x)(-x)] = \ln x^2$. By the rules of logarithms the left side of this equation is equal to $2\ln(-x)$, while the right side is $2\ln x$; so we get, after canceling the 2, $\ln(-x) = \ln x$. This "proof" is flawed, however, because it applies the rules of ordinary (that is, real-valued) algebra to the domain of complex numbers, for which these rules do not necessarily hold. (It is reminiscent of the "proof" that $i^2 = 1$ instead of -1: $i^2 = (\sqrt{-1}) \cdot (\sqrt{-1}) = \sqrt{[(-1) \cdot (-1)]} = \sqrt{1} = 1$. The error is in the second step, because the rule $\sqrt{a} \cdot \sqrt{b} = \sqrt{(ab)}$ is valid only when the numbers under the radical sign are positive.) In 1747 Euler wrote to D'Alembert and pointed out that a logarithm of a negative number must be complex and, moreover, that *it has infinitely many different values*. Indeed, if x is a negative number, its polar representation is $|x|$ cis π, so that from equation 8 we get $\ln x = \ln|x| + i(\pi + 2k\pi)$, $k = 0, \pm1, \pm2, \ldots$. In particular, for $x = -1$ we have $\ln|x| = \ln 1 = 0$, so that $\ln(-1) = i(\pi + 2k\pi) = i(2k+1)\pi = \ldots, -3\pi i, -\pi i, \pi i, 3\pi i, \ldots$. The principal value of $\ln(-1)$ (the value for $k = 0$) is thus πi, a result that also follows directly from Euler's formula $e^{\pi i} = -1$. The logarithm of an *imaginary* number can be found similarly; for example, since the polar form of $z = i$ is $1 \cdot \mathrm{cis}\,\pi/2$, we have $\ln i = \ln 1 + i(\pi/2 + 2k\pi) = 0 + (2k + 1/2)\pi i = \ldots, -3\pi i/2, \pi i/2, 5\pi i/2, \ldots$.

Needless to say, in Euler's time such results were regarded as strange curiosities. Although by then complex numbers had been fully accepted into the domain of algebra, their application to transcendental functions was still a novelty. It was Euler who broke the ground by showing that complex numbers can be used as an "input" to transcendental functions, provided the "output" is also regarded as a complex number. His new approach produced quite unexpected results. Thus, he showed that *imaginary powers of an imaginary number can be real*. Consider, for example, the expression i^i. What meaning can we give to such an expression? In the first place, a power of any base can always be written as a power of the base e by using the identity

$$b^z = e^{z \ln b} \tag{9}$$

(this identity can be verified by taking the natural logarithm of both sides and noting that $\ln e = 1$). Applying equation 9 to the expression i^i, we have

$$i^i = e^{i \ln i} = e^{i \cdot i(\pi/2 + 2k\pi)} = e^{-(\pi/2 + 2k\pi)}, \quad k = 0, \pm 1, \pm 2, \ldots . \tag{10}$$

We thus get infinitely many values—all of them real—the first few of which (beginning with $k = 0$ and counting backward) are $e^{-\pi/2} \approx 0.208$, $e^{+3\pi/2} \approx 111.318$, $e^{+7\pi/2} \approx 59609.742$, and so on. In a very literal sense, Euler made the imaginary become real![8]

There were other consequences of Euler's pioneering work with complex functions. We saw in Chapter 13 how Euler's formula $e^{ix} = \cos x + i \sin x$ leads to new definitions of the trigonometric functions, $\cos x = (e^{ix} + e^{-ix})/2$ and $\sin x = (e^{ix} - e^{-ix})/2i$. Why not take these definitions and simply replace in them the real variable x by the complex variable z? This would give us formal expressions for the *trigonometric functions of a complex variable*:

$$\cos z = \frac{e^{iz} + e^{-iz}}{2}, \quad \sin z = \frac{e^{iz} - e^{-iz}}{2i}. \tag{11}$$

Of course, in order to be able to calculate the values of $\cos z$ and $\sin z$ for any complex number z, we need to find the real and imaginary parts of these functions. Equation 2 allows us to express both e^{iz} and e^{-iz} in terms of their real and imaginary parts: $e^{iz} = e^{i(x+iy)} = e^{-y+ix} = e^{-y}(\cos x + i \sin x)$ and similarly $e^{-iz} = e^{y}(\cos x - i \sin x)$. Substituting these expressions into equations 11, we get, after a little algebraic manipulation, the formulas

$$\cos z = \cos x \cosh y - i \sin x \sinh y$$

and $$\tag{12}$$

$$\sin z = \sin x \cosh y + i \cos x \sinh y$$

where cosh and sinh denote the hyperbolic functions (see p. 144). One can show that these formulas obey all the familiar properties of the good old trigonometric functions of a real variable. For example, the formulas $\sin^2 x + \cos^2 x = 1$, $d(\sin x)/dx = \cos x$, $d(\cos x)/dx = -\sin x$, and the various addition formulas all remain valid when the real variable x is replaced by the complex variable $z = x + iy$.

An interesting special case of equations 12 arises when we let z be purely imaginary, that is, when $x = 0$. We then have $z = iy$, and equations 12 become

$$\cos(iy) = \cosh y, \quad \sin(iy) = i\sinh y. \tag{13}$$

Are there hyperbolic tensile structures? I'll bet there are.

These remarkable formulas show that in the realm of complex numbers one can go back and forth freely between the circular and hyperbolic functions, whereas in the real domain one can only note the formal analogies between them. The extension to the complex domain essentially removes the distinction between these two classes of functions.

Not only does the extension of a function to the complex domain preserve all its properties from the real domain, it actually endows the function with new features. Earlier in this chapter we saw that a function $w = f(z)$ of a complex variable can be interpreted as a mapping from the z-plane to the w-plane. One of the most elegant theorems in the theory of functions says that at each point where $f(z)$ is analytic (has a derivative), this mapping is *conformal*, or angle-preserving. By this we mean that if two curves in the z-plane intersect at an angle φ, their image curves in the w-plane also intersect at the angle φ. (The angle of intersection is defined as the angle between the tangent lines to the curves at the point of intersection; see fig. 74.) For example, we saw earlier that the function $w = z^2$ maps the hyperbolas $x^2 - y^2 = c$

And the chief characteristic of tensile structure diagrams are the characteristic & invariant angles.

← 23

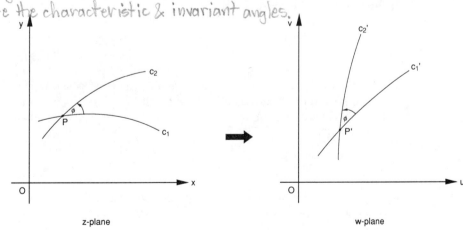

FIG. 74. Conformal property of an analytic function: the angle of intersection of two curves is preserved under the mapping.

and $2xy = k$ onto the lines $u = c$ and $v = k$, respectively. These two families of hyperbolas are *orthogonal*: every hyperbola of one family intersects every hyperbola of the other family at a right angle. This orthogonality is preserved by the mapping, since the image curves $u = c$ and $v = k$ obviously intersect at right angles (see fig. 71). A second example is provided by the function $w = e^z$, which maps the lines $y = c$ and $x = k$ onto the rays $\Phi = c$ and circles $R = e^k$, respectively (fig. 73). Again we see that the angle of intersection—a right angle—is preserved; in this case the conformal property expresses the well-known theorem that every tangent line to a circle is perpendicular to the radius at the point of tangency.

As one might expect, the Cauchy-Riemann equations (equations 1) play a central role in the theory of functions of a complex variable. Not only do they provide the conditions for a function $w = f(z)$ to be analytic at z, but they give rise to one of the most important results of complex analysis. If we differentiate the first of equations 1 with respect to x and the second with respect to y, we get, using Leibniz's notation for the second derivative (with ∂ replacing d; see p. 96),

$$\frac{\partial^2 u}{\partial x^2} = \frac{\partial}{\partial x}\left(\frac{\partial v}{\partial y}\right), \quad \frac{\partial^2 u}{\partial y^2} = -\frac{\partial}{\partial y}\left(\frac{\partial v}{\partial x}\right). \tag{13}$$

The jumble of ∂'s may be confusing, so let us explain: $\partial^2 u/\partial x^2$ is the second derivative of $u(x, y)$ with respect to x, while $\partial/\partial x(\partial v/\partial y)$ is the second "mixed" derivative of $v(x, y)$ with respect to y and x, *in that order*. In other words, we work this expression from the inside outward, just as we do with a pair of nested parentheses [(. . .)]. Similar interpretations hold for the other two expressions. All this seems quite confusing, but fortunately we do not have to worry too much about the order in which we perform the differentiations: if the functions u and v are reasonably "well behaved" (meaning that they are continuous and have continuous derivatives), the order of differentiation is immaterial. That is, $\partial/\partial y(\partial/\partial x) = \partial/\partial x(\partial/\partial y)$—a commutative law of sorts. For example, if $u = 3x^2 y^3$, then $\partial u/\partial x = 3(2x)y^3 = 6xy^3$, $\partial/\partial y(\partial u/\partial x) = 6x(3y^2) = 18xy^2$, $\partial u/\partial y = 3x^2(3y^2) = 9x^2 y^2$, and $\partial/\partial x(\partial u/\partial y) = 9(2x)y^2 = 18xy^2$; hence $\partial/\partial y(\partial u/\partial x) = \partial/\partial x(\partial u/\partial y)$. This result, proved in advanced calculus texts, allows us to conclude that the right sides of equations 13 are equal and opposite, and hence their sum is 0. Thus,

How to create a universe out of nothing

$$\frac{\partial^2 u}{\partial x^2} + \frac{\partial^2 u}{\partial y^2} = 0. \tag{14}$$

A similar result holds for $v(x, y)$. Let us again use the function $w = e^z$ as an example. From equation 2 we have $u = e^x \cos y$, so that $\partial u/\partial x = e^x \cos y$, $\partial^2 u/\partial x^2 = e^x \cos y$, $\partial u/\partial y = -e^x \sin y$, and $\partial^2 u/\partial y^2 = -e^x \cos y$; thus $\partial^2 u/\partial x^2 + \partial^2 u/\partial y^2 = 0$.

Equation 14 is known as Laplace's equation in two dimensions, named for the great French mathematician Pierre Simon Marquis de Laplace (1749–1827). Its generalization to three dimensions, $\partial^2 u/\partial x^2 + \partial^2 u/\partial y^2 + \partial^2 u/\partial z^2 = 0$ (where u is now a function of the three spacial coordinates x, y, and z), is one of the most important equations of mathematical physics. Generally speaking, any physical quantity in a state of equilibrium—an electrostatic field, a fluid in steady-state motion, or the temperature distribution of a body in thermal equilibrium, to name but three examples—is described by the three-dimensional Laplace equation. It may happen, however, that the phenomenon under consideration depends on only two spacial coordinates, say x and y, in which case it will be described by equation 14. For example, we might consider a fluid in steady-state motion whose velocity u is always parallel to the xy plane and is independent of the z-coordinate. Such a motion is essentially two-dimensional. The fact that the real and imaginary parts of an analytic function $w = f(z) = u(x, y) + iv(x, y)$ both satisfy equation 14 means that we can represent the velocity u by the complex function $f(z)$, known as the "complex potential." This has the advantage of allowing us to deal with a single independent variable z, instead of two independent variables x and y. Moreover, we can use the properties of complex functions to facilitate the mathematical treatment of the phenomenon under consideration. We can, for instance, transform the region in the z-plane in which the flow takes place to a simpler region in the w-plane by a suitable conformal mapping, solve the problem there, and then use the inverse mapping to go back to the z-plane. This technique is routinely used in potential theory.[9]

The theory of functions of a complex variable is one of the three great achievements of nineteenth-century mathematics (the others are abstract algebra and non-Euclidean geometry). It signified an expansion of the differential and integral calculus to realms that would have been unimaginable to Newton and Leibniz. Euler, around 1750, was the pathfinder; Cauchy, Riemann, Weierstrass, and many others in the nineteenth century gave it the status it enjoys today. (Cauchy, incidentally, was the first to give a precise definition of the limit concept, dispensing with the vague notions of fluxions and differentials.) What would have been the reaction of Newton and Leibniz had they lived to see their brainchild grow to maturity? Most likely it would have been one of awe and amazement.

NOTES AND SOURCES

1. Quoted in Robert Edouard Moritz, *On Mathematics and Mathematicians (Memorabilia Mathematica)* (1914; rpt. New York: Dover, 1942), p. 282.

2. For a history of negative and complex numbers, see Morris Kline,

Mathematics: The Loss of Certainty (New York: Oxford University Press, 1980), pp. 114–121, and David Eugene Smith, *History of Mathematics*, 2 vols. (1923; rpt. New York: Dover, 1958), 2:257–260.

3. Gauss in fact gave four different proofs, the last one in 1850. For the second proof, see David Eugene Smith, *A Source Book in Mathematics* (1929; rpt. New York: Dover, 1959), pp. 292–306.

4. The theorem is true even when the polynomial has complex coefficients; for example, the polynomial $x^3 - 2(1 + i)x^2 + (1 + 4i)x - 2i$ has the three roots 1, 1, and $2i$.

5. An example of a function for which this condition is not met is the absolute-value function $y = |x|$, whose V-shaped graph forms a 45° angle at the origin. If we attempt to find the derivative of this function at $x = 0$, we get two different results, 1 or –1, depending on whether we let $x \to 0$ from the right or from the left. The function has a "right-sided derivative" at $x = 0$ and a "left-sided derivative" there, but not a single derivative.

6. See any book on the theory of functions of a complex variable.

7. This can be verified by starting with $e^w \cdot e^z$, replacing each factor by the corresponding right side of equation 2, and using the addition formulas for sine and cosine.

8. More on the debate regarding logarithms of negative and imaginary numbers can be found in Florian Cajori, *A History of Mathematics* (1894), 2d ed. (New York: Macmillan, 1919), pp. 235–237.

9. However, this can be done only in two dimensions. In three dimensions other methods must be used, for example, vector calculus. See Erwin Kreyszig, *Advanced Engineering Mathematics* (New York: John Wiley, 1979), pp. 551–558 and ch. 18.

15

But What Kind of Number Is It?

Number rules the universe.

—MOTTO OF THE PYTHAGOREANS

The history of π goes back to ancient times; that of e spans only about four centuries. The number π originated with a problem in geometry: how to find the circumference and area of a circle. The origins of e are less clear; they seem to go back to the sixteenth century, when it was noticed that the expression $(1 + 1/n)^n$ appearing in the formula for compound interest tends to a certain limit—about 2.71828—as n increases. Thus e became the first number to be *defined* by a limiting process, $e = \lim(1 + 1/n)^n$ as $n \to \infty$. For a while the new number was regarded as a kind of curiosity; then Saint-Vincent's successful quadrature of the hyperbola brought the logarithmic function and the number e to the forefront of mathematics. The crucial step came with the invention of calculus, when it turned out that the inverse of the logarithmic function—later to be denoted by e^x—is equal to its own derivative. This at once gave the number e and the function e^x a pivotal role in analysis. Then around 1750 Euler allowed the variable x to assume imaginary and even complex values, paving the way to the theory of functions of complex variables, with their remarkable properties. One question, however, still remained unanswered: Exactly what kind of number is e?

From the dawn of recorded history humans have had to deal with numbers. To the ancients—and to some tribes even today—numbers meant the counting numbers. Indeed, so long as one needs only to take stock of one's possessions, the counting numbers (also called *natural numbers* or *positive integers*) are sufficient. Sooner or later, however, one must deal with measurement—to find the area of a tract of land, or the volume of a flask of wine, or the distance from one town to another. And it is highly unlikely that such a measurement will result in an exact number of units. Thus the need for fractions.

Fractions were already known to the Egyptians and Babylonians, who devised ingenious ways to record them and compute with them. But it was the Greeks, influenced by the teachings of Pythagoras, who made fractions the central pillar of their mathematical and philo-sophical system, elevating them to an almost mythical status. The Pythagoreans believed that everything in our world—from physics and cosmology to art and architecture—can be expressed in terms of fractions, that is, rational numbers. This belief most likely originated with Pythagoras' interest in the laws of musical harmony. He is said to have experimented with various sound-producing objects—strings, bells, and glasses filled with water—and discovered a quanti-tative relation between the length of a vibrating string and the pitch of the sound it produces: the shorter the string, the higher the pitch. Moreover, he found that the common musical intervals (distances between notes on the musical staff) correspond to simple *ratios* of string lengths. For example, an octave corresponds to a length ratio of $2:1$, a fifth to a ratio of $3:2$, a fourth to $4:3$, and so on (the terms *octave*, *fifth*, and *fourth* refer to the positions of these intervals in the musical scale; see p. 129). It was on the basis of these ratios—the three "perfect intervals"—that Pythagoras devised his famous mu-sical scale. But he went further. He interpreted his discovery to mean that not only is musical harmony ruled by simple ratios of integers but so is the entire universe. This extraordinary stretch of logic can be understood only if we remember that in Greek philosophy music—and more precisely, the *theory* of music (as opposed to mere perfor-mance)—ranked equal in status to the natural sciences, particularly mathematics. Thus, Pythagoras reasoned that if music is based on rational numbers, surely the entire universe must be too. Rational numbers thus dominated the Greek view of the world, just as rational thinking dominated their philosophy (indeed, the Greek word for ra-tional is *logos*, from which the modern word *logic* derives).

Very little is known about Pythagoras' life; what we do know comes entirely from works written several centuries after his death, in which reference is made to his discoveries. Hence, almost every-thing said about him must be taken with a good deal of skepticism.[1] He was born around 570 B.C. on the island of Samos in the Aegean Sea. Not far from Samos, in the town of Miletus on the mainland of Asia Minor, lived Thales, the first of the great Greek philosophers. It is thus quite possible that young Pythagoras—fifty years Thales' jun-ior—went to Miletus to study under the great scholar. He then trav-eled throughout the ancient world and eventually settled in the town of Crotona, in what is now southern Italy, where he founded his fa-mous school of philosophy. The Pythagorean school was more than just a forum for philosophical discussions; it was a mystic order whose members were bound by strict laws of secrecy. The Pythag-oreans kept no written records of their discussions. But what they discussed had an enormous influence on Europe's scientific thinking well into the Renaissance. One of the last Pythagoreans was the great astronomer Johannes Kepler (1571–1630), whose ardent belief in the dominance of rational numbers led him astray for more than thirty years in his search for the laws of planetary motion.

It is, of course, not only philosophical arguments that make the rational numbers so central to mathematics. One property that distinguishes these numbers from the integers is this: the rationals form a *dense* set of numbers. By this we mean that between any two fractions, no matter how close, we can always squeeze another. Take the fractions 1/1,001 and 1/1,000 as an example. These fractions are certainly close, their difference being about one-millionth. Yet we can easily find a fraction that lies between them, for example, 2/2,001. We can then repeat the process and find a fraction betwen 2/2,001 and 1/1,000 (for example, 4/4,001), and so on ad infinitum. Not only is there room for another fraction between any two given fractions, there is room for *infinitely* many new fractions. Consequently, we can express the outcome of any measurement in terms of rational numbers alone. This is because the accuracy of any measurement is inherently limited by the accuracy of our measuring device; all we can hope for is to arrive at an approximate figure, for which rational numbers are entirely sufficient.

The word *dense* accurately reflects the way the rationals are distributed along the number line. Take any segment on the line, no matter how small: it is always populated by infinitely many "rational points" (that is, points whose distances from the origin are given by rational numbers). So it seems only natural to conclude—as the Greeks did—that the entire number line is populated by rational points. But in mathematics, what *seems* to be a natural conclusion often turns out to be false. One of the most momentous events in the history of mathematics was the discovery that the rational numbers, despite their density, leave "holes" along the number line—points that do not correspond to rational numbers.

The discovery of these holes is attributed to Pythagoras, though it may well have been one of his disciples who actually made it; we shall never know, for out of deference to their great master the Pythagoreans credited all their discoveries to him. The discovery involved the diagonal of a unit square (a square whose side is equal to 1). Let us denote the length of the diagonal by x; by the Pythagorean Theorem we have $x^2 = 1^2 + 1^2 = 2$, so that x is the square root of 2, written $\sqrt{2}$. The Pythagoreans, of course, assumed that this number is equal to some fraction, and they desperately tried to find it. But one day one of them made the startling discovery that $\sqrt{2}$ cannot equal a fraction. Thus the existence of *irrational numbers* was discovered.

In all likelihood, the Greeks used a geometric argument to demonstrate that $\sqrt{2}$ is irrational. Today we know of several nongeometric proofs of the irrationality of $\sqrt{2}$, all of them "indirect" proofs. We start from the assumption that $\sqrt{2}$ *is* a ratio of two integers, say m/n, and then show that this assumption leads to a contradiction, and that consequently $\sqrt{2}$ cannot equal the supposed ratio. We assume that m/n is in lowest terms (that is, m and n have no common factors). Here the various proofs go in different directions. We may, for example, square the equation $\sqrt{2} = m/n$ and get $2 = m^2/n^2$, hence $m^2 = 2n^2$. This means that m^2, and therefore m itself, is an even integer (because the square of an odd integer is always odd). Thus $m = 2r$ for some integer r. We then have $(2r)^2 = 2n^2$, or, after simplifying, $n^2 = 2r^2$. But this means that n, too, is even, so $n = 2s$. Thus both m and n are even integers and have therefore the common factor 2, contrary to our assumption that the fraction m/n is in lowest terms. Therefore $\sqrt{2}$ cannot be a fraction. Clever.

The discovery that $\sqrt{2}$ is irrational left the Pythagoreans in a state of shock, for here was a quantity that could clearly be measured and even constructed with a straightedge and compass, yet it was not a rational number. So great was their bewilderment that they refused to think of $\sqrt{2}$ as a number at all, in effect regarding the diagonal of a square as a numberless magnitude! (This distinction between arithmetic number and geometric magnitude, which in effect contradicted the Phythagorean doctrine that number rules the universe, would henceforth become an essential element of Greek mathematics.) True to their pledge of secrecy, the Pythagoreans vowed to keep the discovery to themselves. But legend has it that one of them, a man named Hippasus, resolved to go his own way and reveal to the world the existence of irrational numbers. Alarmed by this breach of loyalty, his fellows conspired to throw him overboard the ship they were sailing on. from

But knowledge of the discovery spread, and soon other irrational numbers were found. For example, the square root of every prime number is irrational, as are the square roots of most composite numbers. By the time Euclid compiled his *Elements* in the third century B.C., the novelty of irrational numbers had by and large faded. Book X of the *Elements* gives an extensive geometric theory of irrationals, or *incommensurables*, as they were called—line segments with no common measure. (If the segments AB and CD had a common measure, their lengths would be exact multiples of a third segment PQ; we would thus have $AB = mPQ$, $CD = nPQ$ for some integers m and n, hence $AB/CD = (mPQ)/(nPQ) = m/n$, a rational number.) A fully satisfactory theory of irrationals, however—one devoid of geometric considerations—was not given until 1872, when Richard Dedekind (1831–1916) published his celebrated essay *Continuity and Irrational Numbers*.

If we unite the set of rational numbers with the irrationals we get the larger set of *real numbers*. A real number is any number that can be written as a decimal. These decimals are of three types: terminating, such as 1.4; nonterminating and repeating, such as 0.2727 ...

$0.\overline{27}$

(also written as $0.\overline{27}$); and nonterminating, nonrepeating, such as 0.1010010001 ..., where the digits never recur in exactly the same order. It is well known that decimals of the first two types always represent rational numbers (in the examples given, 1.4 = 7/5 and 0.2727 ... = 3/11), while decimals of the third type represent irrational numbers.

The decimal representation of real numbers at once confirms what we said earlier: from a practical point of view—for the purpose of measurement—we do not need irrational numbers. For we can always approximate an irrational number by a series of *rational approximations* whose accuracy can be made as good as we wish. For example, the sequence of rational numbers 1, 1.4 (= 7/5), 1.41 (= 141/100), 1.414 (= 707/500), and 1.4142 (= 7,071/5,000) are all rational approximations of $\sqrt{2}$, progressively increasing in accuracy. It is the *theoretical* aspects of irrational numbers that make them so important in mathematics: they are needed to fill the "holes" left on the number line by the existence of nonrational points; they make the set of real numbers a complete system, a *number continuum*.

Matters thus stood for the next two and a half millennia. Then, around 1850, a new kind of number was discovered. Most of the numbers we encounter in elementary algebra can be thought of as solutions of simple equations; more specifically, they are solutions of polynomial equations with integer coefficients. For example, the numbers −1, 2/3, and $\sqrt{2}$ are solutions of the polynomial equations $x + 1 = 0$, $3x − 2 = 0$, and $x^2 − 2 = 0$, respectively. (The number $i = \sqrt{-1}$ also belongs to this group, since it satisfies the equation $x^2 + 1 = 0$; we will, however, confine our discussion here to real numbers only.) Even a complicated-looking number such as $\sqrt[3]{(1 − \sqrt{2})}$ belongs to this class, since it satisfies the equation $x^6 − 2x^3 − 1 = 0$, as can easily be checked. A real number that satisfies (is a solution of) a polynomial equation with integer coefficients is called *algebraic*. } definition

Clearly every rational number a/b is algebraic, since it satisfies the equation $bx − a = 0$. Thus if a number is *not* algebraic, it must be irrational. The converse, however, is not true: an irrational number may be algebraic, as the example of $\sqrt{2}$ shows. The question therefore arises: Are there any *nonalgebraic* irrational numbers? By the beginning of the nineteenth century mathematicians began to suspect that the answer is yes, but no such number had actually been found. It seemed that a nonalgebraic number, if ever discovered, would be an oddity.

It was in 1844 that the French mathematician Joseph Liouville (1809–1882) proved that nonalgebraic numbers do indeed exist. His proof, though not simple,[2] enabled him to produce several examples of such numbers. One of his examples, known as Liouville's number, was

Holy murgatroyd. $\frac{1}{10^{1!}} + \frac{1}{10^{2!}} + \frac{1}{10^{3!}} + \frac{1}{10^{4!}} + \cdots,$

whose decimal expansion is 0.110001000000000000000000100 ...
(the increasingly long blocks of zeros are due to the presence of $n!$
in the exponent of each denominator in Liouville's number, causing the terms to decrease extremely fast). Another example is
0.1234567891011121314 ..., where the digits are the natural numbers
in order. A real number that is not algebraic is called *transcendental.*
There is nothing mystic about this word; it merely indicates that these
numbers transcend (go beyond) the realm of algebraic numbers.

In contrast to the irrational numbers, whose discovery arose from
a mundane problem in geometry, the first transcendental numbers
were created specifically for the purpose of demonstrating that such
numbers exist; in a sense they were "artificial" numbers. But once
this goal was achieved, attention turned to some more commonplace
numbers, specifically π and e. That these two numbers are *irrational*
had been known for more than a century: Euler in 1737 proved the
irrationality of both e and e^2,[3] and Johann Heinrich Lambert (1728–
1777), a Swiss-German mathematician, proved the same for π in
1768.[4] Lambert showed that the functions e^x and $\tan x$ (the ratio
$\sin x/\cos x$) cannot assume rational values if x is a rational number
other than 0.[5] However, since $\tan \pi/4 = \tan 45° = 1$, a rational number, it follows that $\pi/4$ and therefore π must be irrational. Lambert
suspected that π and e are transcendental but could not prove it.

From then on, the stories of π and e became closely intertwined.
Liouville himself proved that e cannot be the solution of a *quadratic*
equation with integer coefficients. But this, of course, falls short of
proving that e is transcendental—that it is not the solution of *any*
polynomial equation with integer coefficients. This task was left to
another French mathematician, Charles Hermite (1822–1901).

Hermite was born with a defect in his leg, a handicap that turned to
his advantage, for it made him unfit for military service. Although his
performance as a student at the prestigious École Polytechnique was
not brilliant, he soon proved himself one of the most original mathematicians of the second half of the nineteenth century. His work covered a wide range of areas, including number theory, algebra, and
analysis (his specialty was elliptic functions, a topic in higher analysis), and his broad outlook enabled him to find many connections
between these seemingly distinct fields. Besides his research, he
wrote several mathematics textbooks that became classics. His famous proof of the transcendence of e was published in 1873 in a
memoir of more than thirty pages. In it Hermite actually gave two

distinct proofs, of which the second was the more rigorous.[6] As a sequel to his proof, Hermite gave the following rational approximations for e and e^2:

$$e \approx \frac{58{,}291}{21{,}444}, \quad e^2 \approx \frac{158{,}452}{21{,}444}.$$

very pretty, for some reason I'm not quite sure of.

The former has the decimal value 2.718289498, in error of less than 0.0003 percent of the true value.

Having settled the status of e, Hermite might have been expected to devote all his efforts to doing the same for π. But in a letter to a former student he wrote: "I shall risk nothing on an attempt to prove the transcendence of π. If others undertake this enterprise, no one will be happier than I in their success. But believe me, it will not fail to cost them some effort."[7] Clearly, he expected the task to be a formidable one. But in 1882, only nine years after Hermite's proof of the transcendence of e, success rewarded the efforts of the German mathematician Carl Louis Ferdinand Lindemann (1852–1939). Lindemann modeled his proof after that of Hermite; he showed that an expression of the form

$$A_1 e^{a_1} + A_2 e^{a_2} + \ldots + A_n e^{a_n},$$

where the a_i's are distinct algebraic numbers (real or complex) and the A_i's are algebraic numbers, can never be 0 (we exclude the trivial case where all the A_i's are 0).[8] But we know one such expression that *is* equal to 0: Euler's formula $e^{\pi i} + 1 = 0$ (note that the left side can be written as $e^{\pi i} + e^0$, which has the required form). Therefore πi, and hence π, cannot be algebraic: π is transcendental.

With these developments, the long inquiry into the nature of the circle ratio came to a conclusion. The transcendence of π settled once and for all the age-old problem of constructing, by straightedge and compass alone, a square equal in area to a given circle. This celebrated problem had obsessed mathematicians ever since Plato, in the [*fascist!*] third century B.C., decreed that all geometric constructions should be accomplished with only a straightedge (an unmarked ruler) and a compass. It is well known that such a construction can be done only if the lengths of all the line segments involved satisfy a certain type of polynomial equation with integer coefficients.[9] Now the area of a circle of unit radius is π; so if this area is to equal the area of a square of side x, we must have $x^2 = \pi$ and hence $x = \sqrt{\pi}$. But to construct a segment of this length, $\sqrt{\pi}$ and therefore π must satisfy a polynomial equation with integer coefficients, making it an algebraic number. Since π is not algebraic, the construction is impossible.

$A = \pi r^2$ if $r = 1$ then $A = \pi$, period.

The solution of a mystery that had puzzled mathematicians since antiquity made Lindemann famous. Yet it was Hermite's proof of the transcendence of e that paved the way for Lindemann's proof. In comparing the contributions of the two mathematicians, the *Dictionary of Scientific Biography* has this to say: "Thus Lindemann, a mediocre mathematician, became even more famous than Hermite for a discovery for which Hermite had laid all the groundwork and that he had come within a gnat's eye of making."[10] Later in his life Lindemann attempted to solve another famous problem, Fermat's Last Theorem, but his proof was found to have a serious error at the very beginning.[11]

mediocre?

Why mediocre?

Poor Lindemann.

In one sense the stories of π and e differ. Because of the longer history and greater fame of π, the urge to compute it to an ever greater number of digits has over the years become something of a race. Even Lindemann's proof that π is transcendental has not stopped the digit hunters from performing ever more spectacular feats (the record for 1989 was 480 million decimal places). No such craze befell e.[12] Nor has e generated the same amount of trivia as π,[13] although I did find the following footnote in a recent book on physics: "For those familiar with American history, the first nine digits [of e] after the decimal point can be remembered by $e = 2.7$ (Andrew Jackson)^2, or $e = 2.718281828 \ldots$, because Andrew Jackson was elected President of the United States in 1828. For those good in mathematics, on the other hand, this is a good way to remember their American history."[14]

With the nature of the two most famous numbers of mathematics settled, it seemed that the attention of mathematicians would turn to other areas. But at the Second International Congress of Mathematicians, held in Paris in 1900, one of the towering mathematicians of the time, David Hilbert (1862–1943), challenged the mathematical community with a list of twenty-three unsolved problems whose solution he regarded as of the utmost importance. The seventh problem on Hilbert's list was to prove or disprove the hypothesis that for any algebraic number $a \neq 0$, 1 and any irrational algebraic number b, the expression a^b is always transcendental; as specific examples he gave the numbers $2^{\sqrt{2}}$ and e^π (the latter because it can be written as i^{-2i} [see p. 178] and thus has the required form).[15] Hilbert predicted that this problem would take longer to solve than Fermat's Last Theorem, but he was overly pessimistic. In 1929 the Russian mathematician Alexandr Osipovich Gelfond (1906–1968) proved the transcendence of e^π, followed a year later by the proof for $2^{\sqrt{2}}$. Hilbert's general hypothesis regarding a^b was proved in 1934 by Gelfond and independently by T. Schneider of Germany.

Translation from e^π to i^{-2i}.

It is not easy to prove that a specific given number is transcendental: one must prove that the number does *not* fulfill a certain requirement. Among the numbers whose status has not yet been settled are π^e, π^π, and e^e. The case of π^e is particularly interesting, for it reminds

now π^e is a curious number.

broken symmetry?

us of the skewed symmetry that exists between π and e. As we saw in Chapter 10, e plays a role with respect to the hyperbola somewhat similar to that of π with respect to the circle. But this similarity is not perfect, as Euler's formula $e^{\pi i} = -1$ clearly shows (π and e occupy different positions in it). The two famous numbers, despite their close association, have quite different personalities.

The discovery of transcendental numbers did not create the same intellectual shock that irrational numbers had produced twenty-five hundred years earlier, but its consequences were equally significant. It showed that behind the seeming simplicity of the real number system hide many subtleties, subtleties that cannot be discerned by simply looking at the decimal expansion of a number. The biggest surprise was yet to come. In 1874 the German mathematician Georg Cantor (1845–1918) made the startling discovery that there are more irrational numbers than rational ones, and more transcendental numbers than algebraic ones. In other words, far from being oddities, *most* real numbers are irrational; and among irrational numbers, most are transcendental![16]

But this takes us to ever higher realms of abstraction. If we content ourselves with computing the numerical values of π^e and e^π, we find that they are surprisingly close: 22.459157... and 23.140692 ..., respectively. Of course, π and e themselves are not that far apart numerically. Think of it: of the infinity of real numbers, those that are most important to mathematics—0, 1, $\sqrt{2}$, e and π—are located within less than four units on the number line. A remarkable coincidence? A mere detail in the Creator's grand design? I let the reader decide. *No mere coincidence.*

→ And $2 + 2 = 2 \times 2 = 4$
I believe 4 is the only number with this property.

$e^\pi - \pi^e = \begin{array}{r} 23.140692\,... \\ -22.459157... \\ \hline = 0.681535 \end{array}$

The golden section is 1.61803.
See Pp 131

NOTES AND SOURCES

1. See B. L. van der Waerden, *Science Awakening: Egyptian, Babylonian, and Greek Mathematics*, trans. Arnold Dresden (New York: John Wiley, 1963), pp. 92–102.

2. See, for example, George F. Simmons, *Calculus with Analytic Geometry* (New York: McGraw-Hill, 1985), pp. 734–739.

3. A proof of the irrationality of e is given in Appendix 2.

4. Lambert is often credited with introducing hyperbolic functions into mathematics, but Vincenzo Riccati seems to have preceded him (see p. 144).

Very pretty
Very very
pretty!

5. As a result, the exponential curve $y = e^x$ passes through no algebraic points in the plane except the point (0,1). (An algebraic point is a point whose x and y coordinates are both algebraic numbers.) To quote Heinrich Dörrie: "Since algebraic points are omnipresent in densely concentrated quantities within the plane, the exponential curve accomplishes the remarkably difficult feat of winding between all these points without touching any of them. The

a miracle of sorts

24 Sep 1994

same is, naturally, also true of the logarithmic curve $y = \ln x$" (Dörrie, *100 Great Problems of Elementary Mathematics: Their History and Solution*, trans. David Antin [1958; rpt. New York: Dover, 1965], p. 136).

6. See David Eugene Smith, *A Source Book in Mathematics* (1929; rpt. New York: Dover, 1959), pp. 99–106. For Hilbert's simplified version of Hermite's proof, see Simmons, *Calculus with Analytic Geometry*, pp. 737–739.

7. Quoted in Simmons, *Calculus with Analytic Geometry*, p. 843.

8. For a simplified version of Lindemann's proof, see Dörrie, *100 Great Problems*, pp. 128–137.

9. See Richard Courant and Herbert Robbins, *What Is Mathematics?* (1941; rpt. London: Oxford University Press, 1941), pp. 127–140.

10. C. C. Gillispie, editor (New York: Charles Scribner's Sons, 1972).

11. Concerning a recent proof of Fermat's Last Theorem, see Chapter 7, note 1.

12. The poster *Computer e*, by David Slowinski and William Christi (Palo Alto, Calif.: Creative Publications, 1981), shows *e* to 4,030 decimal places. A companion poster, *Computer π*, by Stephen J. Rogowski and Dan Pasco (1979) gives π to 8,182 places.

13. See, for example, Howard Eves, *An Introduction to the History of Mathematics* (1964; rpt. Philadelphia: Saunders College Publishing, 1983), pp. 89 and 97.

14. Edward Teller, Wendy Teller, and Wilson Talley, *Conversations on the Dark Secrets of Physics* (New York and London: Plenum Press, 1991), p. 87.

15. See Ronald Calinger, ed., *Classics of Mathematics* (Oak Park, Ill.: Moore Publishing Company, 1982), pp. 653–677. Hilbert's seventh problem is on p. 667.

16. An account of Cantor's work can be found in my book, *To Infinity and Beyond: A Cultural History of the Infinite* (1987; rpt. Princeton: Princeton University Press, 1991), chs. 9 and 10.

Appendixes

The letter e *may now no longer be used to denote anything other than this positive universal constant* [the solution of the equation ln x = 1]

—EDMUND LANDAU, *Differential and Integral Calculus* (1934)

Appendix 1

Some Additional Remarks on Napier's Logarithms

In his *Mirifici logarithmorum canonis constructio*, published posthumously in 1619, Napier explained his invention of logarithms in terms of a geometric-mechanical model, a common approach to solving mathematical problems at the time (we recall that Newton used a similar model in describing his idea of fluxions). Consider a line segment AB and an infinite ray parallel to AB and extending from C to the right (fig. 75).

FIG. 75. Napier used a geometric model to explain his idea of logarithms: P moves along AB with a speed proportional to the distance PB, while Q moves along CD with a constant speed equal to the initial speed of P. If we put $x = PB$ and $y = CQ$, then y is the (Napierian) logarithm of x.

A point P starts to move from A toward B with a speed that is proportional, at every instant, to the distance from P to B. At the same instant that P starts its motion, a point Q begins to move from C to the right with a *constant* speed equal to the initial speed of P. As time progresses, the distance PB decreases at a rate that is itself decreasing, while the distance CQ increases at a uniform rate. Napier defined the distance of Q from its initial position C as the logarithm of the distance of P from its *final* position B. If we put $PB = x$ and $CQ = y$, we have

$$y = \text{Nap log } x,$$

where Nap log stands for "Napierian logarithm."[1]

We can easily see that this definition indeed transforms a product of two numbers (represented as distances along AB) into a sum of two other numbers (distances from C). Suppose that the segment AB is of unit length, and let us mark off equal segments of arbitrary length along the ray from C; we will label these 0, 1, 2, 3, and so on. Since Q moves at a uniform speed, it will cover these segments in equal time intervals. As P starts to move from A, Q is at 0 (point C); when P is at the half mark of AB, Q is at 1; when P has covered 3/4 of AB, Q is at 2, and so on. Since x measures the distance that P still has to go until it reaches B, we have the following table:

x	1	1/2	1/4	1/8	1/16	1/32	1/64	...
y	0	1	2	3	4	5	6	...

This is actually a very primitive table of logarithms: each number in the lower row is the logarithm (to the base 1/2) of the corresponding number in the upper row. Indeed, the sum of any two numbers in the lower row corresponds to the product of the corresponding numbers in the upper row. Note that in this table y increases with decreasing x, in contrast to our modern (base 10 or base e) logarithms, which increase with increasing numbers.

As we mentioned in Chapter 1, in keeping with the practice in trigonometry to divide the radius of a unit circle into 10,000,000 parts, Napier took the distance AB to be 10^7. If we assume that the initial speed of point P is also 10^7, we can describe the motion of P and Q in terms of the two differential equations $dx/dt = -x$, $dy/dt = 10^7$, with the initial conditions $x(0) = 10^7$, $y(0) = 0$. Eliminating t between these equations, we get $dy/dx = -10^7/x$, whose solution is $y = -10^7 \ln x + c$. Since $y = 0$ when $x = 10^7$, we have $c = 10^7 \ln 10^7$, and thus $y = -10^7 (\ln x - \ln 10^7) = -10^7 \ln (x/10^7)$. Using the formula $\log_b x = -\log_{1/b} x$, we can write the solution as $y = 10^7 \log_{1/e}(x/10^7)$, or $y/10^7 = \log_{1/e}(x/10^7)$. This shows that, apart from the factor 10^7 (which merely amounts to shifting the decimal point), Napier's logarithms are actually logarithms to the base $1/e$, though he himself never thought in terms of a base.[2]

SOURCES

1. Excerpts, with commentary, from Napier's *Constructio* can be found in Ronald Calinger, ed., *Classics of Mathematics* (Oak Park, Ill.: Moore Publishing Company, 1982), pp. 254–260, and in D. J. Struik, ed., *A Source Book in Mathematics, 1200–1800* (Cambridge, Mass.: Harvard University Press, 1969), pp. 11–21. See also the facsimile edition of Wright's 1616 English translation of Napier's *Descriptio*: John Nepair, *A Description of the Admirable Table of Logarithms* (Amsterdam: Da Capo Press, 1969), ch. 1.

2. Carl B. Boyer, *A History of Mathematics*, rev. ed. (1968; rpt. New York: John Wiley, 1989), pp. 349–350.

Appendix 2

The Existence of $\lim (1 + 1/n)^n$ as $n \to \infty$

We first show that the sequence

$$S_n = 1 + \frac{1}{1!} + \frac{1}{2!} + \ldots + \frac{1}{n!}, \quad n = 1, 2, 3, \ldots$$

converges to a limit as n increases without bound. This sum increases with each additional term, so we have $S_n < S_{n+1}$ for all n; that is, the sequence S_n increases monotonically. Beginning with $n = 3$, we also have $n! = 1 \cdot 2 \cdot 3 \cdot \ldots \cdot n > 1 \cdot 2 \cdot 2 \cdot \ldots \cdot 2 = 2^{n-1}$; therefore

$$S_n < 1 + 1 + \frac{1}{2} + \frac{1}{2^2} + \ldots + \frac{1}{2^{n-1}}$$

for $n = 3, 4, 5, \ldots$. Now, in this last sum the terms starting with the second form a geometric progression with the common ratio 1/2. The sum of this progression is $(1 - 1/2^n)/(1 - 1/2) = 2(1 - 1/2^n) < 2$. We therefore have $S_n < 1 + 2 = 3$, showing that the sequence S_n is bounded from above by 3 (that is, the values of S_n never exceed 3). We now use a well-known theorem from analysis: Every bounded, monotone increasing sequence tends to a limit as $n \to \infty$. Thus S_n converges to a limit S. Our proof also shows that S is between 2 and 3.

We now consider the sequence $T_n = (1 + 1/n)^n$. We will show that this sequence converges to the same limit as S_n. By the binomial theorem,

$$T_n = 1 + n \cdot \frac{1}{n} + \frac{n(n-1)}{2!} \cdot \frac{1}{n^2} + \ldots + \frac{n(n-1)(n-2) \ldots 1}{n!} \cdot \frac{1}{n^n}$$

$$= 1 + 1 + \left(1 - \frac{1}{n}\right) \cdot \frac{1}{2!} + \ldots$$

$$+ \left(1 - \frac{1}{n}\right)\left(1 - \frac{2}{n}\right) \ldots \left(1 - \frac{n-1}{n}\right) \cdot \frac{1}{n!}$$

Since the expression within each pair of parentheses is less than 1, we have $T_n \leq S_n$ (actually, $T_n < S_n$ beginning with $n = 2$). Therefore the sequence T_n is also bounded from above. Moreover, T_n increases monotonically, because replacing n with $n + 1$ only causes the sum to increase. Thus T_n too converges to a limit as $n \to \infty$. We denote this limit by T.

We now show that $S = T$. Since $S_n \geq T_n$ for all n, we have $S \geq T$. We will show that at the same time $S \leq T$. Let $m < n$ be a fixed integer. The first $m + 1$ terms of T_n are

$$1 + 1 + \left(1 - \frac{1}{n}\right) \cdot \frac{1}{2!} + \dots$$
$$+ \left(1 - \frac{1}{n}\right)\left(1 - \frac{2}{n}\right) \dots \left(1 - \frac{m-1}{n}\right) \cdot \frac{1}{m!}$$

Because $m < n$ and all terms are positive, this last sum is less than T_n. If we now let n increase without bound while holding m fixed, the sum will tend to S_m, while T_n will tend to T. We thus have $S_m \leq T$, and consequently $S \leq T$. Since we have already shown that $S \geq T$, it follows that $S = T$, which is what we wished to prove. The limit T, of course, is the number e.

As a sequel, we prove that e is irrational.[1] Our proof is indirect: we assume that e is *rational* and then show that this assumption leads to a contradiction. Let $e = p/q$, where p and q are integers. We already know that $2 < e < 3$, so e cannot be an integer; consequently the denominator q must be at least 2. We now multiply both sides of the equation

$$e = 1 + \frac{1}{1!} + \frac{1}{2!} + \frac{1}{3!} + \dots + \frac{1}{n!} + \dots$$

by $q! = 1 \cdot 2 \cdot 3 \cdot \dots \cdot q$. On the left side this gives us

$$e \cdot q! = \left(\frac{p}{q}\right) \cdot 1 \cdot 2 \cdot 3 \cdot \dots \cdot q = p \cdot 1 \cdot 2 \cdot 3 \cdot \dots \cdot (q-1)$$

while on the right side we get

$$[q! + q! + 3 \cdot 4 \cdot \dots \cdot q + 4 \cdot 5 \cdot \dots \cdot q + \dots$$
$$+ (q-1) \cdot q + q + 1] + \frac{1}{q+1} + \frac{1}{(q+1)(q+2)} + \dots$$

(note that the 1 inside the brackets comes from the term $1/q!$ in the series for e). The left side is obviously an integer, because it is the product of integers. On the right side, the expression inside the brackets is likewise an integer. But the remaining terms are not integers, because each denominator is at least 3. We now show that their sum, too, is not an integer. Since $q \geq 2$, we have

$$\frac{1}{q+1} + \frac{1}{(q+1)(q+2)} + \dots \leq \frac{1}{3} + \frac{1}{3 \cdot 4} + \dots$$
$$< \frac{1}{3} + \frac{1}{3^2} + \frac{1}{3^3} + \dots = \frac{1}{3} \cdot \frac{1}{1 - \frac{1}{3}} = \frac{1}{2},$$

where we used the formula for the sum of an infinite geometric series, $a + ar + ar^2 + \ldots = a/(1 - r)$, for $|r| < 1$. Thus we have an integer on the left side of the equation and a non-integer on the right side, obviously a contradiction. Hence e cannot be the ratio of two integers—it is irrational.

SOURCE

1. Richard Courant and Herbert Robbins, *What Is Mathematics?* (1941; rpt. London: Oxford University Press, 1969), pp. 298–299.

Appendix 3

A Heuristic Derivation of the
Fundamental Theorem of Calculus

In figure 76, let A be the area under the graph of a function $y = f(x)$ from some fixed value of x, say $x = a$ (called "the lower limit of integration") to a variable value ("the upper limit"). To avoid confusion, let us denote the upper limit of integration by t, reserving the letter x for the independent variable of the function $f(x)$. The area A then becomes a function of this upper limit: $A = A(t)$. We wish to show that $dA/dt = f(t)$; that is, *the rate of change of the area function $A(t)$ with respect to t is equal to the value of $f(x)$ at $x = t$.*

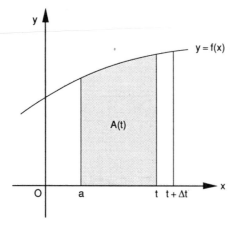

FIG. 76. The Fundamental Theorem of Calculus: the rate of change of the area function $A(t)$ is equal to the value of $f(x)$ at $x = t$.

Let us move from the point $x = t$ to a neighboring point $x = t + \Delta t$; that is, we give t a small increment Δt. The area thereby increases by the amount $\Delta A = A(t + \Delta t) - A(t)$. The added area, for small Δt, has the approximate shape of a rectangular strip of width Δt and height $y = f(t)$, as can be seen from figure 76. Thus $\Delta A \approx y\Delta t$, with the approximation improving the smaller Δt is. Dividing by Δt, we have $\Delta A/\Delta t \approx y$. Going to the limit as $\Delta t \to 0$, the expression on the left becomes the derivative (rate of change) of A with respect to t, dA/dt. We thus have $dA/dt = y = f(t)$, as we wished to show.

This shows that the area A, considered as a function of t, is an antiderivative, or indefinite integral, of $f(t)$: $A = \int f(t)dt$. To fix the value of A for any particular choice of t, we write $A = {}_a\int^t f(x)dx$, where we denoted the variable of integration by x.[1] Note that $\int f(t)dt$ is a *function* (the area function), while ${}_a\int^t f(x)dx$ is a *number*, called the *definite integral of $f(x)$ from $x = a$ to $x = t$.*

Clearly this derivation is not a rigorous proof; for a full proof, see any good calculus text.

NOTE

1. The variable of integration x is a "dummy variable"; it can be replaced by any other letter without affecting the outcome.

Appendix 4

The Inverse Relation between $\lim (b^h - 1)/h = 1$ and $\lim (1 + h)^{1/h} = b$ as $h \to 0$

Our goal is to determine the value of b for which $\lim_{h\to 0}(b^h - 1)/h = 1$ (see p. 101). We start with the expression $(b^h - 1)/h$ for *finite* h and set it equal to 1:

$$\frac{b^h - 1}{h} = 1. \tag{1}$$

Certainly, if this expression is identically equal to 1, then also $\lim_{h\to 0}(b^h - 1)/h = 1$. We now solve equation 1 for b. We do this in two steps. In the first step we get

$$b^h = 1 + h$$

and in the second,

$$b = {}^h\!\sqrt{(1 + h)} = (1 + h)^{1/h} \tag{2}$$

where we replaced the radical sign with a fractional exponent. Now equation 1 expresses b as an implicit function of h; since equations 1 and 2 are equivalent, letting $h \to 0$ will give us the equivalent expressions

$$\lim_{h\to 0} \frac{b^h - 1}{h} = 1 \quad \text{and} \quad b = \lim_{h\to 0}(1 + h)^{1/h}.$$

The last limit is the number e. Thus, to make the expression $\lim_{h\to 0}(b^h - 1)/h$ equal to 1, b must be chosen as $e = 2.71828 \ldots$.

We stress that this is not a complete proof, only an outline.[1] But from a didactic point of view, it is simpler than the traditional approach, where one starts with the *logarithmic* function, finds its derivative—a rather lengthy process—and only then sets the base equal to e (after which one still must revert to the exponential function to show that $d(e^x)/dx = e^x$).

NOTE

1. For a complete discussion, see Edmund Landau, *Differential and Integral Calculus* (New York: Chelsea Publishing Company, 1965), pp. 39–48.

Appendix 5

An Alternative Definition of the Logarithmic Function

The antiderivative of x^n, apart from an additive constant, is $x^{n+1}/(n + 1)$, a formula that holds for all values of n except -1 (see p. 78). The case $n = -1$ had been a mystery until Grégoire Saint-Vincent found that the area under the hyperbola $y = 1/x = x^{-1}$ follows a logarithmic law. We know now that the logarithm involved is the natural logarithm (see p. 107); hence, if we regard this area as a function of its upper limit and denote it by $A(x)$, we have $A(x) = \ln x$. By the Fundamental Theorem of Calculus we have $d(\ln x)/dx = 1/x$, so that $\ln x$ (or, more generally, $\ln x + c$, where c is an arbitrary constant) is an antiderivative of $1/x$.

We could, however, adopt the reverse approach and *define* the natural logarithm as the area under the graph of $y = 1/x$ from, say $x = 1$ to a variable point $x > 1$.[1] Writing this area as an integral, we have

$$A(x) = \int_1^x \frac{dt}{t}, \tag{1}$$

where we denoted the variable of integration by t to avoid confusing it with the upper limit of integration x (we have also written the expression inside the integral as dt/t, instead of the more formal $(1/t)dt$). Note that equation 1 defines A as a function of the upper limit of integration x. We now show that this function has all the properties of the natural logarithmic function.

We first note that $A(1) = 0$. Second, by the Fundamental Theorem of Calculus we have $dA/dx = 1/x$. Third, for any two positive real numbers x and y we have the *addition law* $A(xy) = A(x) + A(y)$. Indeed,

$$A(xy) = \int_1^{xy} \frac{dt}{t} = \int_1^x \frac{dt}{t} + \int_x^{xy} \frac{dt}{t}, \tag{2}$$

where we have split the interval of integration $[1, xy]$ into two subintervals, $[1, x]$ and $[x, xy]$. The first integral on the right side of equation 2, by our definition, is $A(x)$. For the second integral, we make the substitution (change of variable) $u = t/x$; this gives us $du = dt/x$ (note that x is a constant as far as the integration is concerned). Moreover, the lower limit of integration $t = x$ changes to $u = 1$, and the upper limit $t = xy$ changes to $u = y$. We thus have

$$\int_x^{xy} \frac{dt}{t} = \int_1^y \frac{du}{u} = A(y)$$

(we used the fact that t and u are "dummy variables"; see p. 201). This establishes the addition law.

Finally, since the area under the graph of $1/x$ continuously grows as x increases, A is a *monotone increasing* function of x; that is, if $x > y$, then $A(x) > A(y)$. Thus, as x varies from 0 to infinity, $A(x)$ assumes all real values from $-\infty$ to ∞. But this means that there must be a number—we shall call it e—for which the area under the graph is exactly equal to 1: $A(e) = 1$. It is not difficult to show that this number is the limit of $(1 + 1/n)^n$ as $n \to \infty$; that is, e is the same number that we previously defined as $\lim_{n\to\infty}(1 + 1/n)^n$, or $2.71828\ldots$.[2] In short, the function $A(x)$ defined by equation 1 has all the properties of $\ln x$, and we shall identify it with $\ln x$. And since this function is continuous and monotonically increasing, it has an *inverse*, which we call the natural exponential function and denote by e^x.

This approach may seem somewhat artificial; it certainly has the benefit of wisdom at hindsight, since we already know that the function $\ln x$ has the aforementioned properties. This benefit, however, is not always available to us. There are many simple-looking functions whose antiderivatives cannot be expressed in terms of any finite combination of the elementary functions (polynomials and ratios of polynomials, radicals, and trigonometric and exponential functions and their inverses). An example of such a function is the *exponential integral*, the antiderivative of e^{-x}/x. Although the antiderivative does exist, there is no combination of elementary functions whose derivative is equal to e^{-x}/x. Our only recourse is to *define* the antiderivative as an integral, $\int_x^\infty (e^{-t}/t)\, dt$ (where $x > 0$), denoted as $Ei(x)$, and regard it as a new function. One can deduce the properties of this function, tabulate its values, and graph it just as with any ordinary function.[3] In every respect, then, such "higher" functions should be regraded as known.

NOTES

1. If $0 < x < 1$, we will regard the area as negative. However, $A(x)$ is not defined for $x = 0$ or for negative values of x, since the graph of $1/x$ has an infinite discontinuity at $x = 0$.

2. See Richard Courant, *Differential and Integral Calculus*, vol. 1 (London: Blackie and Son, 1956), pp. 167–177.

3. See Murray R. Spiegel, *Mathematical Handbook of Formulas and Tables*, Schaum's Outline Series, (New York: McGraw-Hill, 1968), pp. 183 and 251.

Appendix 6

Two Properties of the Logarithmic Spiral .

We will prove here two properties of the logarithmic spiral mentioned in the text.

1. Every ray through the origin intersects the spiral at the same angle. (It is because of this property that the logarithmic spiral is also known as the *equiangular spiral*.)

To prove this, we will use the conformal property of the function $w = e^z$, where both z and w are complex variables (see Chapter 14). Representing z in rectangular form as $x + iy$ and w in polar form as $w = R \operatorname{cis} \Phi$, we have $R = e^x$ and $\Phi = y$ (ignoring additions of full rotations) (see p. 176). Thus, vertical lines $x = \text{const.}$ of the z-plane are mapped onto circles $R = e^x = \text{const.}$ concentric about the origin of the w-plane, while horizontal lines $y = \text{const.}$ are mapped onto rays $\Phi = \text{const.}$ emanating from the origin of the w-plane. Consider now a point $P(x, y)$ that moves along the straight line $y = kx$ through the origin of the z-plane. Its image point Q in the w-plane has the polar coordinates $R = e^x$, $\Phi = y = kx$. Eliminating x between these equations, we get $R = e^{\Phi/k}$, which is the polar equation of a logarithmic spiral. Thus, as P traverses the line $y = kx$ in the z-plane, its image point Q describes a logarithmic spiral in the w-plane. Since the line $y = kx$ crosses every horizontal line $y = \text{const.}$ of the z-plane at a fixed angle, say α (where $\tan \alpha = k$), its image curve must cross every ray through the origin of the w-plane at that same angle—a consequence of the fact that our mapping is conformal. This completes the proof.

If we write $a = 1/k = 1/\tan \alpha = \cot \alpha$, we can write the equation of the spiral as $R = e^{a\Phi}$. This shows that there is a connection between the constant a (which determines the rate of growth of the spiral) and the angle α: the smaller α, the greater the rate of growth. For $\alpha = 90°$ we have $a = \cot \alpha = 0$ and therefore $R = 1$, the unit circle. The circle is thus a special logarithmic spiral whose rate of growth is 0.

2. The arc length from any point on the logarithmic spiral to the pole (center) is finite, although it takes infinitely many rotations to reach the pole.

We use the formula for the arc length of a curve given in polar form as $r = f(\theta)$:

$$s = \int_{\theta_1}^{\theta_2} \sqrt{r^2 + (\tfrac{dr}{d\theta})^2}\ d\theta.$$

(This formula can be established by considering a small element of arc length ds and using the Pythagorean Theorem: $ds^2 = (dr)^2 + (rd\theta)^2$.) For the logarithmic spiral we have $r = e^{a\theta}$, $dr/d\theta = ae^{a\theta} = ar$. Thus,

$$s = \int_{\theta_1}^{\theta_2} \sqrt{r^2 + (ar)^2}\ d\theta = \sqrt{1 + a^2} \int_{\theta_1}^{\theta_2} e^{a\theta}\ d\theta$$

$$= \frac{\sqrt{1 + a^2}}{a}(e^{a\theta_2} - e^{a\theta_1}). \tag{1}$$

Let us assume that $a > 0$; that is, r increases as we move along the spiral in a counterclockwise sense (a left-handed spiral). Thinking of θ_2 as fixed and letting $\theta_1 \to -\infty$, we have $e^{a\theta_1} \to 0$, and so

$$s_\infty = \lim_{\theta_1 \to -\infty} s = \frac{\sqrt{1 + a^2}}{a} e^{a\theta_2} = \frac{\sqrt{1 + a^2}}{a} r_2. \tag{2}$$

Thus, for a left-handed spiral, the arc length from any point to the pole is given by equation 2, whose right side has a finite value. For a right-handed spiral ($a < 0$), we will let $\theta_1 \to +\infty$, arriving at a similar conclusion.

The expression on the right side of equation 2 can be interpreted geometrically. Substituting $a = \cot\alpha$ in equation 2 and using the trigonometric identities $1 + \cot^2\alpha = 1/\sin^2\alpha$ and $\cot\alpha = \cos\alpha/\sin\alpha$, we find that $[\sqrt{(1 + a^2)}]/a = 1/\cos\alpha$. Hence $s_\infty = r/\cos\alpha$, where we have dropped the subscript 2 under the r. Referring to figure 77 and taking P as the point from which we measure the arc length to the pole, we have $\cos\alpha = OP/PT = r/PT$. Hence $PT = r/\cos\alpha = s_\infty$; that is, the distance along the spiral from P to the pole is equal to the length of the tangent line to the spiral from P to T. This remarkable fact was discovered in 1645 by Evangelista Torricelli, a disciple of Galileo, using the sum of an infinite geometric series to approximate the arc length.

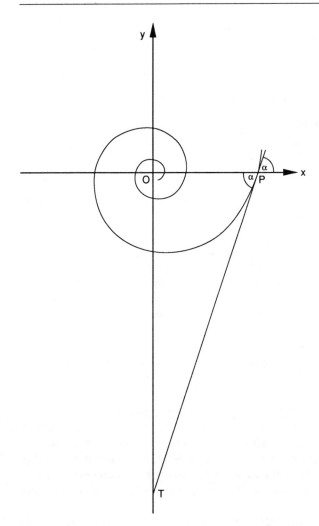

FIG. 77. Rectification of the logarithmic spiral: the distance *PT* is equal to the arc length from *P* to *O*.

Appendix 7

Interpretation of the Parameter φ in the Hyperbolic Functions

The circular or trigonometric functions are defined on the unit circle $x^2 + y^2 = 1$ by the equations

$$\cos\varphi = x, \quad \sin\varphi = y \tag{1}$$

where x and y are the coordinates of a point P on the circle, and φ is the angle between the positive x-axis and the line segment OP, measured counterclockwise in radians. The hyperbolic functions are defined in a similar manner for a point P on the hyperbola $x^2 - y^2 = 1$:

$$\cosh\varphi = x, \quad \sinh\varphi = y. \tag{2}$$

Here the parameter φ cannot be interpreted as an angle. Nevertheless, we can give φ a geometric meaning that will highlight the analogy between the two families of functions.

We first note that the parameter φ in equations 1 can also be thought of as *twice the area of a circular sector of angular width φ and radius 1* (fig. 78). This follows from the formula for the area of a circular sector, $A = r^2\varphi/2$ (note that this formula is valid only if φ is in radians). We will now show that exactly the same meaning can be given to φ in equations 2, where a hyperbolic sector replaces the circular sector.

The shaded area OPR of figure 79 is equal to the difference in the areas of the triangle OPS and the region RPS, where the coordinates of R and S are $(1, 0)$ and $(x, 0)$, respectively. The former area is given by $xy/2$ and the latter by $\int_1^x y\,dx$. Replacing y by $\sqrt{(x^2 - 1)}$ and denoting the variable of integration by t, we thus have

$$A_{OPR} = \frac{x\sqrt{x^2 - 1}}{2} - \int_1^x \sqrt{t^2 - 1}\ dt. \tag{3}$$

To evaluate the integral $\int_1^x \sqrt{(t^2 - 1)}dt$, we make the substitution $t = \cosh u$, $dt = \sinh u\,du$. This changes the interval of integration from $[1, x]$ to $[0, \varphi]$, where $\varphi = \cosh^{-1}x$. If we use the hyperbolic identity $\cosh^2 u - \sinh^2 u = 1$, equation (3) becomes

$$A_{OPR} = \frac{1}{2}\cosh\varphi \sinh\varphi - \int_0^\varphi \sinh^2 u\,du$$

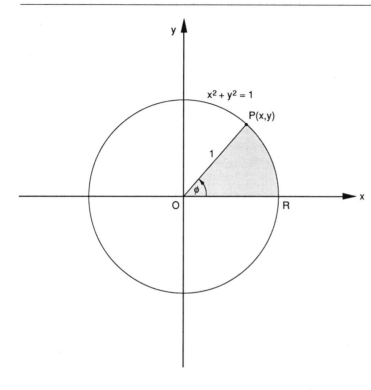

FIG. 78. The unit circle $x^2 + y^2 = 1$. The angle φ can be interpreted as twice the area of the circular segment *OPR*.

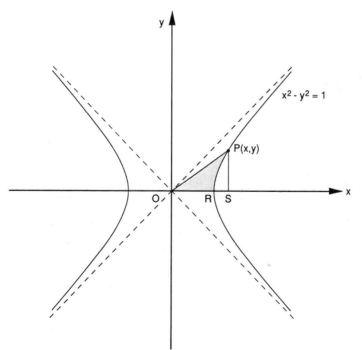

FIG. 79. The rectangular hyperbola $x^2 - y^2 = 1$. If we put $x = \cosh\varphi$, $y = \sinh\varphi$, then the parameter φ can be interpreted as twice the area of the hyperbolic segment *OPR*.

We now use the hyperbolic identities $\sinh 2u = 2 \sinh u \cosh u$ and $\sinh^2 u = (\cosh 2u - 1)/2$. The last equation then becomes

$$A_{OPR} = \frac{1}{4} \sinh 2\varphi - \frac{1}{2} \int_0^\varphi \left(\cosh 2u - 1 \right) du$$

$$= \frac{1}{4} \sinh 2\varphi - \frac{1}{2} \left(\frac{\sinh 2\varphi}{2} - \varphi \right) = \frac{\varphi}{2}.$$

Thus the parameter φ equals *twice the area of the hyperbolic segment OPR*, in exact analogy with the circular functions. As mention earlier, this fact was first noted by Vincenzo Riccati around 1750.

Appendix 8

e to One Hundred Decimal Places

$$e = 2.71828\ 18284\ 59045\ 23536$$
$$02874\ 71352\ 66249\ 77572$$
$$47093\ 69995\ 95749\ 66967$$
$$62772\ 40766\ 30353\ 54759$$
$$45713\ 82178\ 52516\ 64274$$

Source: *Encyclopedic Dictionary of Mathematics*, The Mathematical Society of Japan (Cambridge, Mass.: MIT Press, 1980).

Bibliography

Ball, W. W. Rouse. *A Short Account of the History of Mathematics*. 1908. Rpt. New York: Dover, 1960.

Baron, Margaret E. *The Origins of the Infinitesimal Calculus*. 1969. Rpt. New York: Dover, 1987.

Beckmann, Petr. *A History of π*. Boulder, Colo.: Golem Press, 1977.

Bell, Eric Temple. *Men of Mathematics*, 2 vols. 1937. Rpt. Harmondsworth: Penguin Books, 1965.

Boyer, Carl B. *History of Analytic Geometry: Its Development from the Pyramids to the Heroic Age*. 1956. Rpt. Princeton Junction, N.J.: Scholar's Bookshelf, 1988.

———. *A History of Mathematics* (1968). Rev. ed. New York: John Wiley, 1989.

———. *The History of the Calculus and its Conceptual Development*. New York: Dover, 1959.

Broad, Charlie Dunbar. *Leibniz: An Introduction*. London: Cambridge University Press, 1975.

Burton, David M. *The History of Mathematics: An Introduction*. Boston: Allyn and Bacon, 1985.

Cajori, Florian. *A History of Mathematics* (1893). 2d ed. New York: Macmillan, 1919.

———. *A History of Mathematical Notations*. Vol. 1: *Elementary Mathematics*. Vol. 2, *Higher Mathematics*. 1928–1929. Rpt. La Salle, Ill.: Open Court, 1951.

———. *A History of the Logarithmic Slide Rule and Allied Instruments*. New York: The Engineering News Publishing Company, 1909.

Calinger, Ronald, ed. *Classics of Mathematics*. Oak Park, Ill.: Moore Publishing Company, 1982.

Christianson, Gale E. *In the Presence of the Creation: Isaac Newton and His Times*. New York: Free Press, 1984.

Cook, Theodore Andrea. *The Curves of Life: Being an Account of Spiral Formations and Their Application to Growth in Nature, to Science and to Art*. 1914. Rpt. New York: Dover, 1979.

Coolidge, Julian Lowell. *The Mathematics of Great Amateurs*. 1949. Rpt. New York: Dover, 1963.

Courant, Richard. *Differential and Integral Calculus*, 2 vols. 1934. Rpt. London: Blackie and Son, 1956.

Courant, Richard, and Herbert Robbins. *What Is Mathematics?*. 1941. Rpt. London: Oxford University Press, 1969.

Dantzig, Tobias. *Number: The Language of Science*. 1930. Rpt. New York: Free Press, 1954.

Descartes, René. *La Géométrie* (1637). Trans. David Eugene Smith and Marcia L. Latham. New York: Dover, 1954.

Dörrie, Heinrich. *100 Great Problems of Elementary Mathematics: Their History and Solution*. Trans. David Antin. 1958. Rpt. New York: Dover, 1965.

Edwards, Edward B. *Pattern and Design with Dynamic Symmetry*. 1932. Rpt. New York: Dover, 1967.

Eves, Howard. *An Introduction to the History of Mathematics*. 1964. Rpt. Philadelphia: Saunders College Publishing, 1983.

Fauvel, John, Raymond Flood, Michael Shortland, and Robin Wilson, eds. *Let Newton Be!* New York: Oxford University Press, 1988.

Geiringer, Karl. *The Bach Family: Seven Generations of Creative Genius*. London: Allen and Unwin, 1954.

Ghyka, Matila. *The Geometry of Art and Life*. 1946. Rpt. New York: Dover, 1977.

Gillispie, Charles Coulston, ed. *Dictionary of Scientific Biography*. 16 vols. New York: Charles Scribner's Sons, 1970–1980.

Gjersten, Derek. *The Newton Handbook*. London: Routledge and Kegan Paul, 1986.

Hall, A. R. *Philosophers at War: The Quarrel between Newton and Leibniz*. Cambridge: Cambridge University Press, 1980.

Hambidge, Jay. *The Elements of Dynamic Symmetry*. 1926. Rpt. New York: Dover, 1967.

Heath, Thomas L. *The Works of Archimedes*. 1897; with supplement, 1912. Rpt. New York: Dover, 1953.

Hollingdale, Stuart. *Makers of Mathematics*. Harmondsworth: Penguin Books, 1989.

Horsburgh, E. M., ed. *Handbook of the Napier Tercentenary Celebration, or Modern Instruments and Methods of Calculation*. 1914. Rpt. Los Angeles: Tomash Publishers, 1982.

Huntley, H. E. *The Divine Proportion: A Study in Mathematical Beauty*. New York: Dover, 1970.

Klein, Felix. *Famous Problems of Elementary Geometry* (1895). Trans. Wooster Woodruff Beman and David Eugene Smith. New York: Dover, 1956.

Kline, Morris. *Mathematical Thought from Ancient to Modern Times*. New York: Oxford University Press, 1972.

———. *Mathematics: The Loss of Certainty*. New York: Oxford University Press, 1980.

Knopp, Konrad. *Elements of the Theory of Functions*. Trans. Frederick Bagemihl. New York: Dover, 1952.

Knott, Cargill Gilston, ed. *Napier Tercentenary Memorial Volume*. London: Longmans, Green and Company, 1915.

Koestler, Arthur. *The Watershed: A Biography of Johannes Kepler*. 1959. Rpt. New York: Doubleday, Anchor Books, 1960.

Kramer, Edna E. *The Nature and Growth of Modern Mathematics*. 1970. Rpt. Princeton: Princeton University Press, 1981.

Lützen, Jesper. *Joseph Liouville, 1809–1882: Master of Pure and Applied Mathematics*. New York: Springer-Verlag, 1990.

MacDonnell, Joseph, S.J. *Jesuit Geometers*. St. Louis: Institute of Jesuit Sources, and Vatican City: Vatican Observatory Publications, 1989.

Manuel, Frank E. *A Portrait of Issac Newton*. Cambridge, Mass.: Harvard University Press, 1968.

Maor, Eli. *To Infinity and Beyond: A Cultural History of the Infinite*. 1987. Rpt. Princeton: Princeton University Press, 1991.

Nepair, John. *A Description of the Admirable Table of Logarithms*. Trans. Edward Wright. [London, 1616]. Facsimile ed. Amsterdam: Da Capo Press, 1969.

Neugebauer, Otto. *The Exact Sciences in Antiquity*. 2d ed., 1957. Rpt. New York: Dover, 1969.

Pedoe, Dan. *Geometry and the Liberal Arts*. New York: St. Martin's, 1976.

Runion, Garth E. *The Golden Section and Related Curiosa*. Glenview, Ill.: Scott, Foresman and Company, 1972.

Sanford, Vera. *A Short History of Mathematics*. 1930. Cambridge, Mass.: Houghton Mifflin, 1958.

Simmons, George F. *Calculus with Analytic Geometry*. New York: McGraw-Hill, 1985.

Smith, David Eugene. *History of Mathematics*. Vol. 1: *General Survey of the History of Elementary Mathematics*. Vol. 2: *Special Topics of Elementary Mathematics*. 1923. Rpt. New York: Dover, 1958.

————. *A Source Book in Mathematics*. 1929. Rpt. New York: Dover, 1959.

Struik, D. J., ed. *A Source Book in Mathematics, 1200–1800*. Cambridge, Mass.: Harvard University Press, 1969.

Taylor, C. A. *The Physics of Musical Sounds*. London: English Universities Press, 1965.

Thompson, D'Arcy W. *On Growth and Form*. 1917. Rpt. London and New York: Cambridge University Press, 1961.

Thompson, J. E. *A Manual of the Slide Rule: Its History, Principle and Operation*. 1930. Rpt. New York: Van Nostrand Company, 1944.

Toeplitz, Otto. *The Calculus: A Genetic Approach*. Trans. Luise Lange. 1949. Rpt. Chicago: University of Chicago Press, 1981.

Truesdell, C. *The Rational Mechanics of Flexible or Elastic Bodies, 1638–1788*. Switzerland: Orell Füssli Turici, 1960.

Turnbull, H. W. *The Mathematical Discoveries of Newton*. London: Blackie and Son, 1945.

van der Waerden, B. L. *Science Awakening* (1954). Trans. Arnold Dresden. 1961. Rpt. New York: John Wiley, 1963.

Wells, David. *The Penguin Dictionary of Curious and Interesting Numbers*. Harmondsworth: Penguin Books, 1986.

Westfall, Richard S. *Never at Rest: A Biography of Isaac Newton*. Cambridge: Cambridge University Press, 1980.

Whiteside, D. T., ed. *The Mathematical Papers of Isaac Newton*. 8 vols. Cambridge: Cambridge University Press, 1967–1984.

Yates, Robert C. *Curves and Their Properties*. 1952. Rpt. Reston, Va.: National Council of Teachers of Mathematics, 1974.

Index